T0137259

Computational Social Sciences

Computational Social Sciences

A series of authored and edited monographs that utilize quantitative and computational methods to model, analyze and interpret large-scale social phenomena. Titles within the series contain methods and practices that test and develop theories of complex social processes through bottom-up modeling of social interactions. Of particular interest is the study of the co-evolution of modern communication technology and social behavior and norms, in connection with emerging issues such as trust, risk, security and privacy in novel socio-technical environments. Computational Social Sciences is explicitly transdisciplinary: quantitative methods from fields such as dynamical systems, artificial intelligence, network theory, agent-based modeling, and statistical mechanics are invoked and combined with state-of-the-art mining and analysis of large data sets to help us understand social agents, their interactions on and offline, and the effect of these interactions at the macro level. Topics include, but are not limited to social networks and media, dynamics of opinions, cultures and conflicts, socio-technical co-evolution and social psychology. Computational Social Sciences will also publish monographs and selected edited contributions from specialized conferences and workshops specifically aimed at communicating new findings to a large transdisciplinary audience. A fundamental goal of the series is to provide a single forum within which commonalities and differences in the workings of this field may be discerned, hence leading to deeper insight and understanding.

Federico Cecconi
Editor

AI in the Financial Markets

New Algorithms and Solutions

 Springer

Editor
Federico Cecconi (ID)
Laboratory of Agent Based Social
Simulation
Institute of Cognitive Sciences
and Technologies-National Research
Council
Rome, Italy

ISSN 2509-9574 ISSN 2509-9582 (electronic)
Computational Social Sciences
ISBN 978-3-031-26520-4 ISBN 978-3-031-26518-1 (eBook)
https://doi.org/10.1007/978-3-031-26518-1

This Springer imprint is published by the registered company Springer Nature Switzerland AG
The registered company address is: Gewerbestrasse 11, 6330 Cham, Switzerland

Preface

Financial markets are likely one of the few human achievements that can truly be described as complex systems. Complex systems are structures in physics that: (a) derive a significant portion of their dynamics from interactions between components, (b) the interactions are highly nonlinear and tend to change based on their own dynamics (feedback), (c) the system's behavior cannot be directly attributed to a pure sum of the individual interactions: The total is much more than the sum of the individual parts, (d) and from this derive two very important consequences: a very strong dependence on initial conditions (starting from similar situations, we observe completely divergent final states) (a typical example is weather forecasts).

Financial markets are all of these things, and as such, they present enormous challenges to those attempting to comprehend trends, balances, crisis moments, and high variability.

Then there's AI, or artificial intelligence, and the technology that's been dubbed "the oil of the future," powering a large portion of our socioeconomic systems. AI can be found on the Internet, and in applications, we use every day, in economic forecasting systems, advertising management systems, and search engines. In our cars, games, and, of course, financial markets.

This book is about what happens (and what might happen in the near future) when AI technologies become more prevalent in the capital markets. This is a significant advancement over, say, high-frequency trading algorithms, which raged unchecked in markets (particularly financial derivatives) until a few years ago. These algorithms are gradually being supplanted by extremely complex AI technologies capable of performing tasks that are deeply integrated with the ecosystem in which markets operate (the world of news, or sociopolitical dynamics, the pandemic, geopolitical tensions).

The book is divided into two parts: the theory, which is what AI is today, the current financial markets, and what could be the largest socioeconomic policy operation ever attempted in history, the so-called Great Reset. Then there are the applications, which stem from the use of mixed big data/AI technologies for the detection and deactivation of fake news, economic evaluation and prediction systems, models based on artificial agents capable of making independent decisions, and dynamic opinion models.

All of this is to answer the question: Can artificial intelligence make financial markets safer, more reliable, and more profitable? And what are the dangers?

Thanks

This book was made possible by the resources provided by QBT Sagl https://qbt.ch/, a Swiss company that has been working in the AI field for many years, specifically in the fintech and proptech sectors. QBT provided the model development and simulation data collection hardware and software resources.

Rome, Italy   Federico Cecconi
                                                     federico.cecconi@istc.cnr.it

Contents

About the Editor

Federico Cecconi is R&D manager for QBT Sagl (https://www.qbt.ch/it/), responsible for the LABSS (CNR) of the computer network management and computational resources for simulation, and consultant for Arcipelago Software Srl. He develops computational and mathematical models for the LABSS's issues (social dynamics, reputation, normative dynamics). For LABSS performs both the dissemination that training. His current research interest is twofold: on the one hand, the study of socio-economic phenomena, typically using computational models and databases. The second is the development of AI model for fintech and proptech.

Chapter 1
Artificial Intelligence and Financial Markets

Federico Cecconi ⓘ

Abstract Artificial intelligence is a branch of computer science concerned with the development of intelligent machines that function and behave similarly to humans. A.I. is based on the notion that a machine can simulate the human brain and that, given enough data, a machine can learn to think and act like a human. As a result, we can apply artificial intelligence to financial markets and predict future trends, for example. A machine can also be trained to recognize patterns in buyer dynamics and then use this knowledge to forecast future prices. I could go on and on, but the main point is: why is AI useful in the context of credit markets? What are the advantages of applying artificial intelligence to finance? In brief, this book is dedicated to explaining this: Improved accuracy: AI can aid in the prediction of closing prices, opening prices, and other financial data. This can lead to better decisions and more profitable trades. Improved service: Artificial intelligence can be used to improve customer service in financial institutions. It can, for example, be used to assist customers with account inquiries or to provide a technical analysis of the market. Reduced costs: AI can assist in lowering the costs of running a financial institution. It can be used to automate processes or improve operational efficiency, for example.

Keywords Artificial intelligence · Financial markets · PropTech · FinTech

1.1 What About AI is Useful for the Financial Markets?

Artificial intelligence is becoming more prevalent in both models and applications. It is expanding in assisted (or autonomous) driving solutions, computer vision, automatic text analysis, chat-bot creation, economic planning applications, and marketing support platforms. But how much of this growth is being reinvested profitably in the financial markets? How much AI can we currently see in the capital market, and, more importantly, how much can we expect to see in the near future? What are the challenges? What technologies are you using?

F. Cecconi (✉)
LABSS-ISTC-CNR, Via Palestro 32, 00185 Rome, Italy
e-mail: federico.cecconi@istc.cnr.it

© The Author(s), under exclusive license to Springer Nature Switzerland AG 2023
F. Cecconi (ed.), *AI in the Financial Markets*, Computational Social Sciences,
https://doi.org/10.1007/978-3-031-26518-1_1

To begin answering these questions, let us define AI today. Artificial intelligence is still a long way from creating an intelligent mind. Indeed, there are numerous indications that this is no longer a hot topic. On the contrary, we are far ahead in the field of so-called restricted artificial intelligence, which, when applied to very specific contexts, ensures excellent results that are superior to human skills and abilities. We must not believe that a commercial chat-bot (such as SIRI) was created to mimic human intelligence and perhaps better understand human cognitive processes. In fact, this ambition, to create a positive feedback loop between understanding the mind, artificial mind realizations, and a new understanding of the mind, does not appear to be on the agenda in either basic research or applications. SIRI is an example of access to distributed databases that uses natural language as an interface, and it has far more in common with traditional IT than with the theory of cognitive systems for searching for information, such as short term memory structures in the human mind.

It is true that there are some research centers around the world that deal with mind and AI in an integrated manner, frequently linking the functioning of the mind to the fact that the mind is connected to a brain and a body in nature. Cognitive models of multisensory integration in humans have been developed and applied to model human body experience over the last few decades. Recent research indicates that Bayesian and connectionist models may drive developments in various branches of robotics: assistive robotic devices may adapt to their human users, e.g., in prosthetics, and humanoid robots may be endowed with human-like capabilities regarding their surrounding space, such as keeping safe or socially appropriate distances from other agents. However, we expect to see AI outside of laboratories beyond the time horizon set by this book (ten years).

In conclusion, despite the fact that AI models have limited abstraction capacity and learn differently than humans, we can now develop systems capable of performing actions and tasks efficiently and effectively (think of the difference with which they learn to recognize a cat a child rather than a computer vision system). However, it is undeniable that in certain areas and with specific tasks, systems utilizing artificial intelligence techniques can produce results that are superior to those of humans.

Returning to the financial markets, it is becoming clear which of these technologies we currently expect to find, which we believe will be discovered soon, and which we believe are consistent with the desire to make markets safer, more profitable, and more connected to society and macroeconomic dynamics. I've written a chapter about each of them.

1.1.1 Automatic Evaluation and PropTech

Humans have always attempted to assign a monetary value to things, particularly since our species has been able to generate a "surplus" of resources that can be conserved and possibly exchanged. In fact, it makes no sense to try to assign a value to things in hunter-gatherer societies, such as the Native American populations of the north, because the things themselves, whatever they are, are used now and immediately and replaced with the new.

Capital markets are the most advanced form of this process because an attempt is made to assign value to financial assets that are sometimes very distant from the 'physical' good that produced them. Let's start with mortgages. Even if we ignore any financially complex mortgage transaction (for example, the consolidation of multiple mortgages into a single object on which to borrow money), a mortgage is a loan that I make using the value of a physical asset as collateral (1 property) that follows value dynamics that are very different from the dynamics of the loan financial object. The property's value may increase or decrease depending on structural changes in the environment in which it is rented (a new metro stop has opened in my neighborhood), whereas the mortgage is influenced, for example, by macroeconomic dynamics linked to interest rates (the bank has less interest in providing mortgages because interest rates tend to be low).

After all, there are dozens of circularities that connect the two values, the financial one linked to the mortgage as a financial object in the strict sense, and the property understood as a physical asset in the strict sense: for example, if macroeconomic dynamics push interest rates lower, and banks pay less, the value of a house can fall, because demand falls (many people do not have enough money to buy and, given the difficulty in obtaining mortgag-es, they give up).

Precisely in the field of real estate, and with the idea of having tools capable of calculating the value of things, AI has made an impressive entrance, creating a large number of models that fall into the proptech macro-category.

The term proptech, which derives from the crasis of the words "property" and "technology," was coined in 2014 in the United Kingdom and refers to solutions, technologies, and tools for the innovation of processes, products, services, and the real estate market. Proptech refers to all of the digital solutions that are revolutionizing various aspects of real estate, such as real estate sales, risk analysis, and financial forecasting models, asset management, and the financing dynamics of development projects. Proptech is, in fact, a minor component of the larger digital transformation of the real estate industry. It is associated with a shift in mindset in the real estate industry and its users regarding technological innovation in data collection, transactions, and building and city design. But there's more to it than just the sale. According to some studies, the demand for rental housing units is rapidly increasing, to the point where we refer to "rent generation."

This shift from property ownership to property rental is a major driver of the proptech trend. Tenants today are constantly looking for better and newer spaces with cutting-edge technology. Consumers want an experience they can control—control of lights, temperature, access, and more—from their mobile phones or other channels, whether it's their work environment, their favorite shops, or the spaces of their daily life. digital. AI systems can help calculate costs, evaluate solutions, and occasionally provide occupants with a digital experience, but with the use of technologies such as augmented/virtual reality, it can dramatically improve the customer experience. Companies can thus offer flexible rentals while also providing a high-quality furnished space experience. Furthermore, fintech-proptech firms can provide a streamlined tenant onboarding process.

Proptech solutions almost always include AI elements, which are almost always based on models that can learn on a regular basis using databases. There are elements of computer vision and natural language understanding in the models at times, particularly when it comes to the possibility of including the ability to react to news that is not necessarily of an economic nature (the typical example is news related to climatic dynamics).

1.1.2 'News Based' Decision Model

We already have systems in operation that read news (even in real time) and are able to suggest actions, or carry them out themselves. The journalistic sectors most predisposed to automation are those that have the ability to access large, well-structured data sets. This is the case, for example, with sports news, weather forecasts, earthquake warnings, traffic and crime news, and of course it is the case with financial news. Here are some examples.

Sport

As for the sports sector, in 2016 the Swedish local media company Östgöta Media launched the "Klackspark" website with the aim of covering every local football match. In Sweden football is very popular, each district has its own team and therefore for the newspapers it is complex to deal with all the matches of the national and international football divisions (men and women).

To be able to have greater coverage, Klackspark decided to use automated software writing algorithms, which collect structured data on each game and automatically publish a hundred-word report on the challenge. It is not a difficult task or that requires particular imagination as what is told is based on who scored the goals, the results of previous matches and the ranking. The short articles produced are then reviewed by reporters who add interesting features.

Additionally, the machine alerts reporters when there may be an interesting story to deal with, such as a player scoring multiple goals in the same match. This allows you to interview the direct interviewee by adding information to the article and in fact making it more juicy.

Meteorology

Can automation improve the quality of our life? This may be the case with the weather news industry which produces automatic bulletins for each city. The speed and accuracy of these reports are essential for this sector. A related case is that of bots that report and issue warnings about earthquakes, having a strong influence on people's lives. A special case was the Los Angeles Times project, the "Quake Bot," which can automatically generate and publish earthquake reports to help readers of possible risks to their existence or loved ones.

The system works by drawing information from the United States Geological Survey (USGS) database. However, in June 2017 the Quake Bot released a bogus report due to an error and issued an alert that turned out to be wrong. It was later discovered that the error was caused by an erroneous design decision. Algorithms are not neutral and objective decision makers but rather opaque, both to protect the companies that created them and use them and for the technical needs of secure system management.

The creators, during the programming phase, must however take into consideration the use that a user will make of the information received to provide a sophisticated product. Speed, large-scale data analysis and accuracy are the reasons for choosing automated journalism. The most important benefit for newsrooms is the opportunity to automate repetitive routine tasks that are not stimulating for journalists. In this way they can use the time for more creative jobs and where critical judgment mindset is required. Only some sectors benefit more than others from automation and therefore it would not make sense to automate everything.

Finance

I chose to report on a financial instrument created by Bloomberg, an international news agency. For more than a decade, the agency has been automatically generating written press releases for its terminal. In 2018, Bloomberg achieved a quarter of its content production through automation (Mandeep et al. 2022). The agency's innovation laboratory (called BHIVE) presented "The Bulletin" the same year, a tool capable of providing an automated summary of news.

The research team concentrated on user preferences, particularly their home page, and how to get them the most up-to-date information in order to provide them with a clear and quick-to-read overview. The Bulletin includes a synopsis of the three most important and current events from the Bloomberg global network. Its goal is to provide information to users who have limited time and only look at their phones for brief periods of time but want to stay fully informed about the world. Artificial intelligence intervenes by sifting through the agency's network of 2700 journalists and analysts to extract the most important details. It is based on natural language processing (NLP), which scans articles for key words and semantics and synthesizes them into a simple sentence ready for use by the user (Ashkan 2022).

1.1.3 Trend Follower

Simply put, AI trading software is designed to analyze stock and trading patterns in real-time. From its analysis of the market, the software can come up with stock recommendations as well as real-time data. The type of information generated by these types of software is highly valuable, as you can use them to optimize your entry and exit into stock positions to maximize profits (Yan 2022; Yao et al. 2022; Gu et al. 2021).

Since trading primarily hinges on making timely decisions based on future price movements in the market, the ability to analyze and predict price movements is a valuable skill in trading. By analyzing the market and predicting price movements on your behalf, AI trading software offers an unprecedented advantage over other types of trading software. At the core of AI trading software is machine learning, which is a subfield of artificial intelligence. Machine learning software attempt to imitate human thinking and behavior. We use the term "imitate" in the sense that they are designed with the capability to learn from improving through exposure to data sets. In other words, machine learning models are able to learn from their mistakes as and improve over time without human intervention.

Most AI trading software can monitor the price behavior of thousands of stocks simultaneously and around the clock. Given a set of pre-determined parameters, the software can monitor the market for specific markers of price movements. It will then notify you of these impending changes, allowing you to act accordingly.

In addition to monitoring stock price movements, some AI trading software can track your performance over time. In a back-testing process, the software analyzes historical data to give you an idea of how well your strategy worked for a given period. Some software also allows you to test different strategies within simulated trading platform. The software sifts through thousands of stocks simultaneously each day, analyzing technical and fundamentals. It also analyzes hundreds of social media and news websites for reports of company-based movements that can affect stock prices.

The software offers a wide selection of features, including providing signals for entry and exit to maximize your profits. It also features real-time trading simulation, price alerts, risk assessment, building and back-testing for various trading strategies, and automatic trading. It also features a virtual trading assistant that can help you analyze chart patterns as well as ideal entry and exit points in real-time.

One advantage to the software is its ability to filter bad stocks. It does this by analyzing the stock's historical performance in contrast to current market conditions. The software is also very easy to use, making it ideal for both beginner traders and professional traders.

The software will then run the data through a variety of complicated financial and engineering models. These include classification, regression, and so forth. The software then compiles the results in a predictive ranking for stocks and other assets.

Aside from the ability to rank stocks using indicators, the software has other very useful features, some of which are also backed by artificial intelligence. One feature includes a paper trading portfolio, which allows you to test out an investment strategy without using real money.

It also features a market news service. This service compiles relevant financial news and a watchlist feature that keeps track of specific stocks and markets. There's also a market analysis tool that filters out the best stocks for you to invest in and a calendar to keep track of stock performance week-on-week.

Finally, looking for an AI trading software that supports crypto trading can be difficult given the lack of available information.

Automated trading leaves the actual trading to a crypto bot, you don't have to manually input your orders. All you have to do is set your trade parameters beforehand. The system will run a string of simulations and give you a list of the best stocks with the best possible outcomes.

The software also features back-testing, which takes a look at your past trading decisions and market conditions to evaluate your performance. It also features a demo portfolio where you can run scenarios with exchanges without using real money or crypto. It also offers a way to track your crypto trading performance individually or in aggregate across multiple exchanges via a single app.

Another critical advantage to the software is the level of security with which it operates. The software connects with crypto exchanges only after passing a strict authentication process for your security. It also lets you control what the trading bot has access to on your trading platform as part of a risk-mitigation process.

1.2 Pattern Discovering

In highly digitised markets, such as the ones for equities and FX products, AI solutions promise to provide competitive pricing, manage liquidity, optimise and streamline execution. Importantly, AI algorithms deployed in trading can enhance liquidity management and execution of large orders with minimum market impact, by optimising size, duration and order size in a dynamic fashion, based on market conditions (Joshi et al. 2021; Liu 2021; Zeyi and Wang 2021; Sanz et al. 2021).

The use of AI and big data in sentiment analysis to identify themes, trends, and trading signals is augmenting a practice that is not new. Traders have mined news reports and management announcements/commentaries for decades now, seeking to understand the stock price impact of nonfinancial information. Today, text mining and analysis of social media posts and twits or satellite data through the use of NPL algorithms is an example of the application of innovative technologies that can inform trading decisions, as they have the capacity to automate data gathering and analysis and identify persistent patterns or behaviours on a scale that a human cannot process.

What differentiates AI-managed trading with systematic trading is the reinforcement learning and adjustment of the AI model to changing market conditions, when traditional systematic strategies would take longer to adjust parameters due to the heavy human intervention involved. Conventional back-testing strategies based on historical data may fail to deliver good returns in real time as previously identified trends break down. The use of ML models shifts the analysis towards prediction and analysis of trends in real time, for example using 'walk forward' tests instead of back testing. Such tests predict and adapt totrends in real time to reduce over-fitting (or curve fitting) in back tests based on historical data and (Yin et al. 2021).

We starting tak now about algo wheel. An algo wheel is a broad term, encompassing fully automated solutions to mostly trader-directed flow.

An AI-based algo wheel is an automated routing process embedding AI techniques to assign a broker algorithm to orders from a pre-configured list of algorithmic solutions (Bharne et al. 2019). In other words, AI-based algo wheels are models that select the optimal strategy and broker through which to route the order, depending on market conditions and trading objectives/requirements.

Investment firms typically use algo wheels for two reasons; first, to achieve performance gains from improved execution quality; second, to gain workflow efficiency from automating small order flow or normalizing broker algorithms into standardized naming conventions. Market participants argue that algo wheels reduce the trader bias around the selection of the broker and broker's algorithm deployed in the marketplace.

From the historical point of view, the use of AI in pattern discovering has gone through different stages of development and corresponding complexity, adding a layer to traditional algorithmic trading at each step of the process. First-generation algorithms consisted of buy or sell orders with simple parameters, followed by algorithms allowing for dynamic pricing.

Second-generation algorithms deployed strategies to break up large orders and reduce potential market impact, helping obtain better prices (so-called 'execution algos'). Current strategies based on deep neural networks are designed to provide the best order placement and execution style that can minimize market impact (Sirirattanajakarin and Suntisrivaraporn 2019). Deep neural networks mimic the human brain through a set of algorithms designed to recognize patterns, and are less dependent on human intervention to function and learn. The use of such techniques can be beneficial for market makers in enhancing the management of their inventory and reduce the cost of their balance sheet. As the development of AI advances, AI algorithms evolve into automated, computer programmed algorithms that learn from the data input used and rely less on human intervention.

In practice, the more advanced forms of AI today are mostly used to identify signals from 'low informational value' incidents in flow-based trading, which consist of less obvious events, harder to identify and extract value from. Rather than help with speed of execution, AI is actually used to extract signal from noise in data and converts this information into decision about trades. Less advanced algorithms are mostly used for 'high informational events', which consist of news of financial events that are more obvious for all participants to pick up and where execution speed is of the essence.

At this stage of their development, ML–based models are therefore not aiming at front-running trades and profit from speed of action, such as HFT strategies. Instead, they are mostly confined to being used offline, for example for the calibration of algorithm parameters and for improving algorithms' decision logic, rather than for execution purposes. In the future, however, as AI technology advances and is deployed in more use cases, it could amplify the capabilities of traditional algorithmic trading, with implications for financial markets. This is expected to occur when AI techniques start getting deployed more at the execution phase of trades, offering increased capabilities for automated execution of trades and serving the entire chain of action from picking up signal, to devising strategies, and executing them. ML-based algos for execution will allow for the autonomous and dynamic adjustment of their own decision logic while trading. In that case, the requirements already applied for algorithmic trading.

1.3 Virtual Assistants

Business intelligence

There are a plethora of software solutions devoted to so-called business intelligence, or the ability to make economic decisions in business contexts. There are numerous software solutions dedicated to what is known as business intelligence, or the ability to make economic decisions in business contexts. One example is Gravity's products (https://www.gogravity.com/) 's.merchandise One example is https://www.gograv ity.com/. I guess an application in a banking environment.

Banks requisite to familiarize to developing prospects of customers, lessen expenses, avoid damages of trade to quicker initiate-up challengers, and discover innovative means to propagate incomes. Banks' needs are demanding to magnify aspects of obstacles in the practice of increasing buyer budgets triggered by swift progress and wide creation outlines (Mendoza Armenta et al. 2018). Banks often face issues to handle with an expanding capacity of queries related to customer call-centers and client correspondent electronic mail, and it is known that the Bank conventional customer-service model has inadequate financial prudence of balance and adjusts unwell. Consequently, banks are implementing chatbots or "hi-tech personality". These could benefit and provide on the stipulation, automatic assistance, such as allocating with repeatedly enquired interrogations; accomplish financial services; and assist with fiscal applications (Kurshan et al. 2020).

Scam

With the emergence of E-commerce deception or fraud online has greater than before and it is not so much possible to avoid. Recently the United States has reported the detection of fraud 15 times of the concrete deception rate. Artificial Intelligence comes in within reach nowadays. Through the support of investigating statistical facts, the procedure of appliance practices can currently identify the false contract deprived of some information to the humanoid specialists with the advancement of the accurateness of actual period authorizations and reducing wrong failures. Multi-national companies are currently exploring with Artificial Intelligence to identify and crosscheck with unreliable prevention in financial sectors. One of the best use of artificial intelligence applications we can observe is with the MasterCard. If the fraudsters are trying to use some else's MasterCard by stealing information and data the artificial Decision Intelligence technology will analyze the actual data and send immediate notifications in the genuine holder's email or smartphones and all related communicating mediums associated with personalized wallets.

Protecting security

Many of the administrations are attempting to gadget the Artificial Intellect in order to raise the safekeeping for operational dealings and associated services. It is possible if there is processer access which can forecast the unlawful databases precisely.

Expenditure Configuration Forecast

Artificial Intelligence is beneficial for shopper expenditure recognition used by numerous companies and financial service sectors. It will be obliging when the cars are stolen or the account is hacked in order to avoid the deception or shoplifting.

Stock Dealer scheme

A computer system has been trained to predict when to buy or sell shares in order to buy or sell shares in order to maximize the profits when to minimize the losses during the uncertainties and meltdown. Client-side user validation: This can again authenticate or recognize the user and permit the deal to come to pass.

References

Mandeep, A. Agarwal, A. Bhatia, A. Malhi, P. Kaler, H.S. Pannu, Machine learning based explainable financial forecasting (2022), pp. 34–38

A. Safari, Data driven artificial neural network LSTM hybrid predictive model applied for international stock index prediction (2022), pp. 115–120

J. Yan, Smart financial real-time control system implementation based on artificial intelligence and data mining (2022), pp. 248–251

W. Yao, X. Ren, J. Su, An inception network with bottleneck attention module for deep reinforcement learning framework in financial portfolio management (2022), pp. 310–316

F. Gu, Z. Jiang, J. Su, Application of features and neural network to enhance the performance of deep reinforcement learning in portfolio management (2021), pp. 92–97

S.R. Joshi, A. Somani, S. Roy, ReLink: complete-link industrial record linkage over hybrid feature spaces (2021), pp. 2625–2636

Y. Liu, Computer method research on risk control identification system based on deep learning (2021), pp. 561–565

M.A. Zeyi, L. Wang, Identifying the impacts of digital technologies on labor market: a case study in the food service industry (2021), pp. 214

J.L.C. Sanz, Y. Zhu, Toward scalable artificial intelligence in finance (2021), pp. 460–469

X. Yin, D. Yan, A. Almudaifer, S. Yan, Y. Zhou, Forecasting stock prices using stock correlation graph: a graph convolutional network approach (2021), pp. 1–8

P.K. Bharne, S. Prabhune, Stock market prediction using artificial neural networks (2019), pp. 64–68

S. Sirirattanajakarin, B. Suntisrivaraporn, Annotation intent identification toward enhancement of marketing campaign performance (2019), pp. 1–5

S.M. Armenta, S.G. Fern, J.B. Comparan, E.G. Zavala, C.G. Villalobos, Legacy financial transmission rights and their function and role they play during the financial transmission rights auction (2018), pp. 35–40

E. Kurshan, H. Shen, H. Yu, Financial crime and fraud detection using graph computing: application considerations and outlook (2020), pp. 125–130

Federico Cecconi is R&D manager for QBT Sagl (https://www.qbt.ch/it/), responsible for the LABSS (CNR) of the computer network management and computational resources for simulation, and consultant for Arcipelago Software Srl. He develops computational and mathematical models for the Labss's issues (social dynamics, reputation, normative dynamics). For Labss performs both the dissemination that training. His current research interest is twofold: on the one hand, the study of socio-economic phenomena, typically using computational models and databases. The second is the development of AI model for fintech and proptech.

Chapter 2
AI, the Overall Picture

Luca Marconi

Just as electricity transformed almost everything 100 years ago, today I actually have a hard time thinking of an industry that I don't think AI will transform in the next several years. —Andrew Ng

Abstract Nowadays, artificial intelligence (AI) algorithms are being designed, exploited and integrated into a wide variety of software or systems for different and heterogenous application domains. AI is definitively and progressively emerging as transversal and powerful technological paradigm, due to its ability not only to deal with big data and information, but especially because it produces, manages and exploits knowledge. Researchers and scientists are starting to explore, from several perspectives, the different and synergetic ways AI will transform heterogenous business models and every segment of all industries.

Keywords Knowledge representation · Reasoning · AI · Machine learning · Cognitive science

Overall, AI has the potential to generally provide higher quality, greater efficiency, and better outcomes than human experts. This significant and strong potential of AI is emerging in potentially every applicative domain, organizational context or business field. Therefore, the impact of AI on our daily and working life is constantly increasing: intelligent systems are being modelled and interconnected by the application of the powerful technological paradigm of the Internet of Things, autonomous vehicles are being produced and tested in heterogeneous driving conditions and intelligent robots are being designed for a variety of challenges and fields.

Indeed, AI is definitively contributing to the so-called fourth industrial revolution, according to Klaus Schwab, the Founder and Executive Chairman of the World Economic Forum. He clearly stated that AI is a fundamental breakthrough in the

L. Marconi (✉)
Advanced School in Artificial Intelligence, Brain, Mind and Society, Institute of Cognitive Sciences and Technologies, Italian National Research Council, Via San Martino Della Battaglia 44, Rome, Italy
e-mail: luca.marc@hotmail.it

technological scenario, able to foster a deep transformation of the way we live and interact. It suffices to mention that Andrew Ng, former chief scientist at Badu and Cofounder at Coursera, said in a keynote speech at the AI Frontiers conference in 2017 that AI is really the new electricity: a disruptive and pervasive transversal technology, able to support and even empower technologies and processes in potentially any field or domain. Therefore, AI is able to stimulate and reshape the current and future evolution of human society and the comprehension of AI methods and techniques is definitively crucial when it comes to depicting the evolving scenario of this decisive technological paradigm.

In this chapter, we will explore the current technological and methodological scenario of AI by presenting the overall picture and the main macro-challenges, approaches and present steps towards what may happen in the future of AI and society. Specifically, we aim to provide readers with general conceptual instruments to approach both the main symbolic and sub-symbolic AI methods. Finally, we will focus on the Human-AI Interaction perspective, by reviewing to what extent humans and AI actually interact and even collaborate in heterogeneous contexts, so as to depict the current stages towards a hybrid and collective socio-AI systemic intelligence.

2.1 Introduction to AI

Nowadays, Artificial Intelligence (AI) is becoming crucial in the whole society: it is definitively one of the main methodological, technological and algorithmic protagonists of the XXI century. The proliferation of AI methods and tools is definitively there to be seen by everybody: new algorithms are continuously being designed, exploited and integrated into a wide variety of software or systems for different and heterogeneous application domains. Practitioners are aware that AI is something more than just a technological field or methodology. Indeed, AI is concretely and progressively emerging as a transversal and powerful technological paradigm: its power derives from its capacity to foster a widespread transformation of society, due to its ability not only to deal with big data and information, but especially because it produces, manages and exploits knowledge. Researchers and scientists are starting to explore, from several perspectives, the different deep and synergetic ways AI is already stimulating the evolution of heterogeneous business models and industry fields (Haefner et al. 2021; Brynjolfsson and Mcafee 2017; Krogh 2018).

According to DARPA (Launchbury 2017), a general perspective on AI should take into consideration a minimum set of macro-abilities to represent the way it collects, processes and manages information: *perceiving, learning, abstracting* and *reasoning*. These cognitive capabilities are not equally distributed into the whole landscape of AI methods. Indeed, there is still no any single formal and widespread accepted definition of AI. In spite of the current challenging search for such unified definition, the potentialities of the AI methods, models and algorithms, as well as their multiple functions, features and potential impacts, are already affecting the human society. Overall, AI has the potential to generally provide higher quality, greater efficiency, and even better outcomes, compared to human experts, in a wide set of tasks and conditions (Agrawal et al. 2018). This significant and strong effects of AI are emerging in potentially every applicative domain, organizational context or business field. Consequently, both academic studies and empirical experience show the constantly increasing impact of AI on our daily and working life: intelligent systems are being modelled and interconnected by the means of the Internet of Things to connect smart devices in powerful ecosystems, autonomous vehicles are being produced and tested in heterogeneous driving conditions and intelligent robots are being designed for a variety of challenges and fields.

In an even broader perspective, AI is one of the main contributors to the so-called *fourth industrial revolution*, according to Klaus Schwab (2017), the Founder and Executive Chairman of the World Economic Forum. He clearly stated that AI is a fundamental breakthrough in the technological scenario, able to foster a deep transformation of the way we live and interact. Its analysis show that the AI is one of the driving forces of such revolution, with multiple, synergetic and complex interactions among the physical, digital and biological domains. It suffices to mention that Andrew Ng, former chief scientist at Badu and Cofounder at Coursera, said in a keynote speech at the AI Frontiers conference in 2017 that AI is really the new electricity: a disruptive and pervasive transversal technology, able to support and even empower technologies and processes in potentially any field or domain. Thus, AI is not just a mere tool, with a limited, yet powerful effect: on the contrary, its actions and interactions in the surrounding human and non-human ecosystems are directly giving rise to their new states, conditions and emerging behaviors, in a complex socio-technical general perspective. Therefore, AI is able to stimulate and reshape the current and future evolution of human society and the comprehension of AI methods and techniques is definitively crucial when it comes to depicting the evolving scenario of this decisive technological paradigm.

In this chapter, we will explore the current technological and methodological scenario of AI by presenting the overall picture and the main macro-challenges, approaches and present steps towards what may happen in the future of AI and society. In the literature, there are various classification and taxonomy approaches for categorizing AI methods. This chapter is not intended to cover all the many issues and approaches, nor to present a global, integrated and all-embracing review, encompassing the whole scenario of the state-of-the-art. Nevertheless, it is aimed at providing a general overview of the main macro-approaches and paradigms, according to two broad dimensions of AI: *cognition* and *interaction*. Specifically, we aim to provide readers with general conceptual instruments to approach both the main symbolic and sub-symbolic AI methods. Finally, we will focus on the Human-AI Interaction perspective, by reviewing to what extent humans and AI actually interact and even collaborate in heterogeneous contexts, so as to depict the current stages towards a hybrid and collective socio-AI systemic intelligence.

2.2 An Historical Perspective

Despite the fundamental role of AI in the current technological scenario, the initial stages of the design, development and evaluation of AI systems are not actually new in the history. The official birth of the discipline dates back to 1956, when Marvin Minsky and John McCarthy hosted the Dartmouth Summer Research Project on Artificial Intelligence (DSRPAI) at Dartmouth College in New Hampshire. The Dartmouth conference involved some of the main and most important historical "fathers" of AI, including Allen Newell, Herbert Simon, Nathaniel Rochester and Claude Shannon. The primary aim of the Dartmouth event was to create a totally new research area focused on the challenge of designing and building intelligent machines. The subsequent proposal (McCarthy et al. 2006) set the foundations for the analytical study of intelligence in computers and for identifying the first basic concepts for applying relevant methodologies, e.g. information theory, to the study of AI.

Nevertheless, the subsequent stages in the development of the discipline were not as easy as the founding fathers supposed. Despite the further creation of the first algorithms and computer programs, the Minsky's optimistic claim that AI would have soon been able to resemble and emulate human intelligence was not proven, notwithstanding the efforts in that period of feverish activities. Some examples of the results of such years include ELIZA (Weizenbaum 1966), a natural language processing tool somehow anticipating the modern chatbots, and the General Problem Solver (Newell et al. 1959), a logic-based program, with the aspiration to create a universal problem solver, but actually able to solve specific classes of circumscribed problems. The basic approaches exploited in that period mostly relied on logic and explicit knowledge representation, in a totally *symbolic* perspective, with the aim to deal with empirical and psychological data by the means of deductive and linear *if–then* logical paths. Overall, the ambition of the first stages of AI was to quickly and strongly give rise of the so-called *strong AI*, or *Artificial General Intelligence*, namely AI systems totally able to emulate and even substitute the human, natural intelligence. Though, the application of the chosen purely-logical approaches gave rise to intrinsically limited methods, exploiting what DARPA peculiarly called *hand-crafted knowledge* (Launchbury 2017) enabled reasoning capacities for just narrowly selected and pre-defined set of limited problems. Such systems were not endowed with any learning or abstracting capability, thus they did not meet the ambitious expectations of the founding fathers of the discipline.

Consequently, both scholars and governments started to question the initial optimistic wave and conceptual blossoming of AI. In the 70's, the US Congress and the British parliament harshly criticized the previous funds dedicated to the study and design of AI systems. In the same years, the British mathematician James Lighthill published the so-called *Lighthill report* (James et al. 1973), commissioned by the British Science Research Council, after a long discussion within universities and scholars: this report definitively started a hard period for AI, where the optimistic outlook given by the previous AI researchers was abandoned. The Lighthill report collected the major reasons for disappointment, including the limited capabilities of the AI systems to solve real-world problems in generalized settings, though the obtained results into the simulation of circumscribed psychological processes were promising. As a result, the research in the AI field was partly abandoned and several support funds were ended.

Nevertheless, the advent of connectionism (McLeod et al. 1998) in 1986 marked a renewed period of flourishing interest and activity in AI, from a different and somehow opposite perspective. Aside from any ambition to replace human intelligence, the goal of the connectionist approach and philosophy was to study the mental models and the related cognitive and behavioral phenomena as emergent processes deriving from the working actions and interactions of interconnected neural units. Such conceptual shift towards the brain mechanisms somehow anticipated the further neuroscience involvement in AI, as well as the synergy and relationships between AI and *complex systems theory* (San Miguel et al. 2012), where brain and neural systems are actually prominent examples and case studies. The rise of connectionism started the fundamental period of study and design of *artificial neural networks*, trying to resemble the cognitive mechanisms on brain, while basically assuming that knowledge strongly derives from the interaction between the neural system and the outer environment. Even if the foundations of machine learning rooted back to the work by Donald Hebb in the 1940s, with the so-called *Hebbian Learning* theory (Wang and Raj 2017), the most significant pillars for its creation and initial development were stimulated by researchers like Marvin Minksy, James McClelland, David Rumelhart and Frank Rosenblatt, who conceived the *perceptron* (Rosenblatt 1958), namely the first model of neural network, thus proposing a basic method for the further studies in the *back-propagation* training and in *feed-forward* neural networks.

At the same time, the course of history lead to other significant steps for the evolution of AI: first, the foundation of a broad range of Artificial Life approaches, which enlarged the ambition to enable algorithms to resemble biological and ethological phenomena, by the means of models like *swarms, cellular automata* (Langton 1986) and *self-organization*. The application of simulations and computer models to the study of self-organizing interacting agents and systems was another step forward to the current awareness of the need to exploit the many potentialities of the synergy between AI and complex systems, in the present and future broad socio-systemic and human-AI perspective. Moreover, the involvement of neuroscience approaches, as well as the rise of *developmental* and *epigenetic robotics* (Lastname et al. 2001), added relevant contributions to understand biological systems by the means of the integration between neuroscience, developmental psychology and engineering sciences.

Though, it is only since the beginning of the XXI century that AI began to unfold its full potential, thanks to the simultaneous exponential growth in computational resources and computer power, as well as the availability of large and heterogeneous set of data in a wide range of domains. The immense potentialities of the continuously rising algorithms for intelligent systems is currently experiencing a period of great splendour and the development of AI methods and techniques is nowadays fostering unavoidable impacts and upcoming phenomena on the whole society.

2.3 AI Impact on Society

Overall, the impact of AI on society is profoundly deep and inherently multifaceted. Despite its still limited "intelligence", as compared to the human one as a whole, the main power of AI stems from its capability to be applied to potentially every application field. AI represent a real quantum leap respect to other technologies, due to its pervasive, disruptive and technological empowering structure: it has been described as an *enabling technology* (Lee et al. 2019; Lyu 2020), empowering the other technologies by its ability to produce, manage and enrich knowledge. These structural pillars make AI something "big" and "more" than just a circumscribed and mere technological artifact. AI is known to be a technological and methodological paradigm for the evolution of both technology itself and organization. Thus, given the widespread set of applications, affecting our world in many situations and contexts, as well as the growing interactions between AI and humans, it is worth to highlight how the impact of AI in society is not marginal, rather it has the potentialities to radically transform society in a *socio-technical* and *socio-AI* systemic perspective.

This process is still happening, giving rise to significant and unavoidable ethical issues. In spite of the powerful nature of AI, humans cannot escape the need to definitively decide whether they mean to manage it as a tool—rather, an impressive methodological and technological paradigm –, namely as a means to an end, or somehow as the end in itself. In another related perspective, the ethical challenges have always root in the human choice about how they aim to design AI and how they intend to exploit the potentialities, capabilities and structural applications of AI. This choice definitively reflects the human free will: thus, it is certainly a fundamental crossroad that only human intelligence and nature can pass. Nevertheless, the examination of the many and relevant ethical implications of AI in society is not the focus of this chapter.

There is no doubt that the impact of AI in our society is constantly growing and evolving: suffice it to consider that, according to the International Data Corporation (IDC), the global investment on AI and market is expected to break the $500 billion mark in 2023 (IDC 2022). Moreover, the statistics portal Statista forecasts that size of the AI market worldwide will reach over half a trillion U.S. dollars by 2024 (Statista 2022). Consequently, the evolution of the job market is requiring workers and stakeholders to adapt their skills and meta-competences to a quickly transforming scenario. According to the World Economic Forum, more than 60% of children starting school today will work in jobs that still do not exist (World Economic Forum 2017). Moreover, the Oxford Martin School of Economics reports that up to the 47% of all American job functions could be automated within 20 years, resulting in a very different world from the one we know and we operate in Frey and Osborne (2017). It is particularly worth to notice that Gartner identifies AI as a basis for two complementary fundamental trends among the Gartner Top 10 Strategic Technology Trends for 2022 (Gartner 2021): *AI engineering* and *Decision Intelligence*. AI engineering is focused on a complete identification and operationalization of Business Process Automation by the means of synergistic technological and methodological approaches, like AI, Robotic Process Automation (RPA), Business Process Management (BPM), with the aim to detect and automate both business and IT processes. Instead, Decision Intelligence is a systemic approach to improve the organizational decision making by a powerful and pervasive use of both human and artificial intelligence, to model and effectively manage highly complex situations and business conditions by *connected, contextual* and *continuous* decision-making processes.

Thus, the impact of AI in potentially any business field is more and more crucial for the evolution and digital transformation of organizations. AI is affecting both the technological and organizational perspectives and processes, thus reshaping organizations and re-defining the interaction between technologies, management and business. Moreover, the continuous growth of AI methods and algorithms results in a almost daily increasing potential to provide higher quality, greater efficiency, and better outcomes than human domain experts in heterogenous fields and conditions. In the organizational and business framework, that means that AI is prone to assist decision-makers and technicians beyond the scope of humans, thus not as a mere tool but as a decision-making assistance, also exploitable for managerial tasks. As an example, AI-based solutions play important roles in Unilever's talent acquisition process (Marr 2019), in Netflix's decision-making processes related to movie plots, directors, and actors (Westcott Grant 2018), and in Pfizer's drug discovery and scientific development activities (Fleming 2018).

Such impact is today fostered by the recent advances in computational power and resources, the exponential increase in data availability, and new machine-learning techniques. Despite the relatively long history of AI, it is today that we have the concurrent presence of such technological factors, enabling a fast evolution of the application of AI in society. The exponential phenomena we are facing by around the beginning of the XXI century—namely, the exponential growth in both computer power and data availability—and the advancement of informatics in methods, algorithms and infrastructures set up a synergistic system of scientific and technological discoveries, market opportunities and social implications. This virtuous cycle has been carefully recognized by the national and international institutions, resulting in an increasing support to technology-transfer projects, involving companies, start-ups, research centers and universities. The attention towards valuable technology-transfer projects to foster innovation in the countries has been confirmed by many institutional task-forces, work groups and especially recent publications of AI national strategy documents, as in the case of Italy.

Though, let us take a step back, to effectively comprehend why and how AI is having such immense impact on the current society. To deeply understand what is the role of AI in producing, managing and exploiting knowledge, it is essential to start thinking about what intelligence and cognitive abilities are. There is a sort of conceptual progression we cannot avoid to examine, from different complementary perspectives: from data to knowledge, from abilities to intelligence, and, further, from society to socio-AI systems.

2.4   The Cognitive Science Framework

In order to provide a conceptually organized overview of the current scenario of AI, it is fundamental to consider the broader cognitive science perspective and discipline. Indeed, the design, realization and evaluation of effective AI systems, able to strongly encapsulate and exploit cognitive abilities, lead us to question ourselves about the precise nature of *intelligence* as a whole. Cognitive science tries to answer such question by the means of its conceptual framework and categorizations of mental features and activities. The aim of this section is definitively not to provide a comprehensive and complete review of the large panorama of such discipline. Rather, we report a limited set of fundamental concepts that prove to be useful for later understanding and appreciate the further descriptions of the main reported macro-categories of AI systems and methods.

First, we have to take into consideration a triadic conceptual framework, already anticipated at the end of the previous section: from data to knowledge, from abilities to intelligence, and, further, from society to socio-AI systems. While the first pillar of this conceptual architecture is related to the objects of the knowledge representation mechanisms, the second one allows to examine the very different cognitive abilities and the structure of intelligence. Finally, the third one enlarges the considered perspectives towards the interaction between human and artificial cognitive systems, anticipating the essential theme of the Human-AI Interaction in a cognitive science perspective.

In a *cognitivist* approach, the mind and its internal working mechanisms could be represented by a cognitive system, collecting, managing and elaborating information, by the means of a priorly defined internal organization and structure (Garnham 2019). Such structure is somehow similar to a computer, exploiting a set of internal cognitive features and abilities, as well as an inherently limited *information processing* capability throughout its transmission channels. Then, in this perspective, the mental mechanisms are theoretically analogous to the ones of a software or, better, of a complete yet limited electronic brain, collecting and elaborating data and information from external input sources and returning information and knowledge in output, by some sort of *knowledge representation* methodologies. Moreover, the progressive improvement and evolution of the elicited knowledge can benefit from a set of iterative feedback, allowing both to better exploit the collected data and their elaboration by the mind mechanisms. Instead, the further evolution of cognitive science and psychology lead to two complementary perspectives: on the one hand, the analogy between the mind and the computer was deepened, by the means of the paradigms of *modularism* (Fodor 1985) and *connectionism* (Butler 1993); on the other hand, the role of *experience* and *social interaction* gained importance, in a more *constructivist* perspective (Amineh and Asl 2015).

Moreover, the nature of intelligence itself is not fixed and pre-defined in a widely or formally accepted way. Similarly to what happens with AI, lacking a unique official definition in the scientific community, the structure of intelligence as a whole is not totally captured by a circumscribed and comprehensive concept, encompassing all the possible cognitive features and characteristics. Among the possible dichotomies studied by scholars, there are the so-called *fluid* and *cristallized* intelligence (Ziegler et al. 2012*)*, examining how cognitive agents respectively reason over new situations and exploit previously elicited knowledge, the difference between logical and *creative* intelligence, up to the concept of *emotional intelligence* (Salovey and Mayer 1990), affecting even more general psychological and sociological situations and conditions. Even in the clinical practice, it is definitely and widely accepted that the current *psychometric IQ tests,* exploited in diagnostic processes, are inherently limited: indeed, they generally suffer from a set of limitations, related to the difficulty of designing and exploiting a totally *culture-free* test (Gunderson and Siegel 2001), as well as their intrinsic capability to just examine restricted sets of cognitive abilities. Notably, the theory of *multiple intelligences* (Fasko 2001) tries to take into consideration a broader framework, involving several macro-abilities, ranging from *linguistic-verbal*, to *logical-mathematical* and *visual-spatial*, and event to *kinesthetic* and *musical-harmonic* intelligence.

To our scope, it is worth to highlight two main aspects:

1. At present, it is functionally useful to focus on the specific cognitive abilities, which are convenient to be studied for the enhancement of AI systems. The DARPA classification (Launchbury 2017) proves to be effective to encompass the wide range of possible cognitive features, by gather them into four macro-abilities to represent the way a cognitive system collects, processes and manages information: *perceiving*, *learning*, *abstracting* and *reasoning.*
2. It is also functionally convenient to take into consideration two main dimensions, when presenting AI systems and their relation with both the human cognitive ability and society: *cognition*, in the sense sketched beforehand, and *interaction*, intended towards both artificial and human agents, in an extended socio-AI perspective.

Therefore, in the next sections, we will provide the main macro-categories of AI approaches, following the previously presented guidelines. The overview we report is not intended to cover all the possible methods and algorithms in the continuously evolving AI scenario: instead, we aim to show an overview of the macro-categories of approaches, from both the cognitive and socio-interaction perspective. Overall, we will provide readers with general conceptual instruments to approach the AI landscape, as well as to appreciate to what extent humans and AI actually interact and even collaborate in heterogeneous contexts and situations, so as to delineate the current stages towards a hybrid and collective socio-AI systemic intelligence.

Now let us zoom in on a significant dichotomy we have to consider to effectively categorize the current AI approaches and their different potentials: symbolic and subsymbolic AI. By focusing on these two macro-categories, we will acknowledge the relevance of such classification to somehow recap the cognitive characteristics shown, as well as to help the reader appreciate the complementary cognitive potentialities of the reported methods.

2.5 Symbolic and Sub-Symbolic AI

The dichotomy between symbolic and sub-symbolic AI already stems from the dichotomy between *cognitivism* (Garnham 2019) and *connectionism* (Butler 1993): such paradigms, indeed, involve two opposite conceptions of intelligence. On the one hand, cognitivism assumes the mind as a computer, following logical and formal pre-defined rules and mechanisms, as well as manipulating symbols, encapsulating heterogeneous data and information, coming from the external environment, and proceeding through an iterative process of feedback and output results. On the other hand, connectionism takes into consideration the biological nature of intelligence, by depicting a cognitive system based on the parallel involvement of single units—*neurons*—elaborating the information perceived by the means of the interaction among the neurons. While cognitivism derives its perspective by the total analogy between the human mind and an artificial information processing system, working in a purely logical and formalized way, connectionism is based on the idea that the cognitive potential of the mind comes directly from the massive interaction among the interconnected neurons, where intelligence emerges from a distributed elaboration of the information and data collected.

As a step forward, it becomes clear how symbolic and sub-symbolic AI differ. The former deals with *symbols* and logical means of knowledge representation and management, whereas the latter works with *numbers*, non-logical and non-formalized representations of reality. Symbolic AI allows to manipulate the symbols by the means of *logical rules*, formed by predicates, connected by *logical operators*. In this way, all the knowledge admitted in the system can be accurately formalized and constrained by the means of the formal boundaries and operations exploited. Sub-symbolic AI, instead, allows to manipulate the informative elements and factors it deals with by the means of *mathematical operators*, able to manage numbers and data collected from the environment, and fostering further learning processes by the connected neurons. Overall, then, symbolic AI works at a *high-level* perspective, e.g. where the elements involved can be objects and relations among them. Instead, sub-symbolic AI operates at a *low-level*, e.g. where the elements involved can be image pixels, place distances, robot joint torque, sound waves etc.

In the symbolic paradigm to AI, we could represent the different problems as *knowledge graphs* (Zhang et al. 2021; Ji et al. 2021), where *nodes* represent specific objects by the means of logical symbols, and *edges* encode the logical operations to manipulate them. Everything is represented in a *state space*, namely a limited set of possible states allowed to configure a specific problem or situation. The structure is inherently *hierarchical* and *sequential*: starting by a *root* exemplifying the initial situation, the different nodes are connected through several *layers* according to the allowed situations and conditions considered. When all the consented actions have been performed in a state, the final node is a *leaf*, encoding the results of all the actions and procedures leading to it. Significantly, there can be different procedures to get a specific result and reaching the attended state. Consequently, there can be various *paths* that can be followed in the graph, involving heterogeneous nodes and edges, from a given root to a desired leaf. Therefore, symbolic AI could be associated to what DARPA (Launchbury 2017) called the *first wave of AI*, dealing with *handcrafted knowledge*, exploiting reasoning capacities for narrowly circumscribed problems, without any learning or abstracting capability and poorly able to deal with informative uncertainty and complexity.

In the sub-symbolic paradigm to AI, instead, the information collected is managed by the means of *artificial neural networks* (Shanmuganathan 2016), composed by interconnected *neurons* able to manage data and information by the means of "embedded" functions. The neurons are activated according to specific thresholds involving some parameters, then manipulate the informative elements managed and the resulting output information can serve as a basis for the system's learning processes and further analysis. Thus, the whole sub-symbolic AI system works through *parallel distributed processing* mechanisms, where information is elaborated in parallel by the different neurons involved. Therefore, in a totally connectionist and *emergentist* perspective, the sub-symbolic paradigm to AI allows to model mental or behavioral phenomena as the emergent processes directly deriving from the interconnection of localized single units, linked in specific networks. These working mechanisms involve *statistical learning*, where models are trained on even large and heterogeneous datasets, as in the case of *Big Data*, related to specific domains. Such *second wave of AI*—as called by DARPA (Launchbury 2017)—is endowed with perceiving and learning capabilities, while it has poor reasoning ability and it totally lacks abstracting power.

In the further sections, we will present the different approaches reported by considering the two main dimensions of cognition and interaction. Regarding cognition, it is useful (but certainly not all-embracing) to focus on this dichotomy between symbolic and sub-symbolic AI methods to get a general idea of the overall picture of AI. In this chosen framework, a general conceptual map of the presented AI approaches and paradigms is visually represented in the following figure. The two dimensions of *cognition* and *interaction* are exploited to position the different methods and paradigms. Cognition is roughly classified between symbolic and sub-symbolic approaches, encompassing the different cognitive features and abilities examined. Interaction, instead, is functionally classified between human-AI and AI-AI interaction. Without any claim to being exhaustive or exclusive in our approach, we only intend to provide readers a visual sketch to better appreciate the following sections and the subsequent reported methods (Fig. 2.1).

While we will examine all these macro-methods and approaches in the further sections of this chapter, it is already useful to contextualize how such methods are positioned in the conceptual framework of the two main dimensions of cognition and interaction. Regarding cognition, we functionally classify the presented AI methods between the symbolic and sub-symbolic ones: thus, it is quite straightforward to position the methods accordingly. Instead, the dimension of interaction functionally takes into consideration both *human-AI* and *AI-AI* interaction, assuming an extended socio-AI perspective. Given such references, it therefore follows that the axes origin point represents the situation where absolutely no interaction exists, either with human or artificial agents. Hence, the methods are positioned according to the

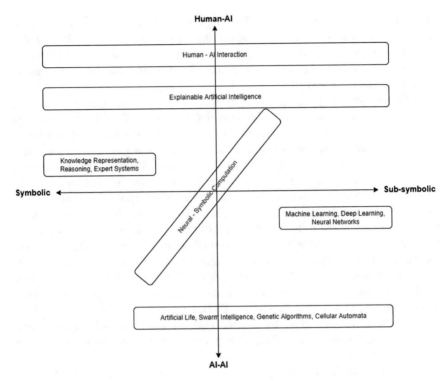

Fig. 2.1 Conceptual map of AI approaches and paradigms

typology and "degree" of interaction in the conceptual map. We recall that this is just a rough visual representation of the reported AI methods, to help the readers better appreciate and comprehend the general overview of the AI landscape, according to the classification methodology provided.

2.6 Knowledge Representation and Reasoning

Let us continue our overview of the current AI scenario by presenting the main and key symbolic methods and approaches, related to the area of knowledge representation. While such methods were highly hyped during the first stages of the AI history, it is undoubtedly clear that they are today relevant and many applications can be found in several application fields. Representing knowledge basically means formalizing it by the means of pre-defined languages, models or methodologies. Thus, the aim of *knowledge representation* (Handbook of Knowledge Representation 2008) is to study, analyze and apply all the widespread frameworks, languages, tools to represent knowledge, as well as to entail different forms of *reasoning* about it. The symbolic perspective is definitively fundamental, due to the key use of *symbols* to represent heterogeneous knowledge bases in different systems. In this perspective, an *intelligent agent* is endowed with a circumscribed set of symbols to represent the outer world (knowledge representation) and is able to make specific *inferences* about it according to *rules* and *principles*. Thus, the main ability of *knowledge-based systems* (KBS) (Akerkar and Sajja 2009; Abdullah et al. 2006) is to logically and symbolically represent an inherently limited part of the world, by the means of a symbolic *modelling language*, as well as to reason, again by the means of a logical and formalized methodology, to discover further characteristics, features or properties of the represented world. In the useful categorization of the AI approaches by DARPA, this exactly corresponds to the *first wave of AI*: systems exploiting *hand-crafted knowledge*, able to even infer significant hidden aspects, features or behaviors of the outer environment or agents in the specific represented world, though they lack fundamental cognitive abilities like *learning* or *abstracting*.

In order to effectively represent knowledge, many methodologies and tools have been widely exploited by researchers, scholars and practitioners. It is particularly worth to recall the use of *logic*, and in particular of *first-order logic* (FOL), as a powerful methodological and formal approach to represent knowledge, adopted in many application fields and domains. FOL is basically able to exploit a wide set of *propositions*, *operators* and *quantifiers* to effectively formalize a *knowledge base* in a pre-defined and specified domain. Moreover, nowadays the role of *fuzzy logic* is fundamental, in order to allow AI systems to represent knowledge and make inferences in highly challenging conditions of informative complexity and uncertainty. Beside logics, it is also important to mention the role of *ontologies* and *knowledge graphs* to capture explicit and naturally circumscribed knowledge in a restricted domain. Indeed, the difference between *explicit knowledge* and *tacit knowledge* proves crucial to comprehend most of the main challenges and reasons of the power and limitations of knowledge representation and KBS. While the former is already expressed in a symbolic framework, thus easy and quite immediate to communicate by some language or system, the latter is *implicit*, empirically acquired by an individual throughout experience, education, tradition, social interactions, intuition and other complex dynamical knowledge-generating processes. Therefore, the use of ontologies and knowledge graphs as means to express conceptual elements, objects, and their relationships, has been traditionally and effectively limited to the representation of explicit knowledge, while recently several proposed methods and algorithms deal with the challenge of formalizing tacit knowledge in both organizational and industrial domains.

In any case, the first and maybe main challenge is before the symbolic formalization phase: indeed, the process of *eliciting knowledge* is crucial and extremely delicate, especially in highly-complex fields or organizational environments. While the KBS are definitively lacking the capability to abstract or extend its reason to unformalized variables, rules or behaviors, the role of the *knowledge engineering* is mainly devoted to improving the quality and quantity of *knowledge management systems* and tools. In complex organizations, this means to enhance or even to revolutionize the whole organizational *information systems* or their current functionalities. Moreover, it also means to make all the involved stakeholders or decision-makers aware of their potential role in the processes of *knowledge sharing*, knowledge management and *knowledge enrichment*. Such knowledge-based improvement of operational processes and human decision-making can be effectively aided by AI tools, which have been studied and applied in a wide range of domains, including heterogeneous operation systems, pharmaceutical and clinical domains, as well as education.

In the field of knowledge representation and reasoning, the most probably known and exploited AI approaches are related to the area of *expert systems*. Starting from the very first wave of AI and right up to the present days, such systems have been extensively designed, developed and applied to a wide range of application environments. The ability of expert systems is definitely to focus on a narrow but precisely represented domain and infer solutions to problems related to that domain. The many applications include clinical and medical ones, industrial contexts and product configurators for sales and marketing. Basically, expert systems are apt to substitute human *domain experts* for specific processes, choices or actions. In this way, they help decision-makers and managers to improve the efficacy and the effectiveness of their decisions and actions in the course of heterogeneous business, organizational or industrial processes: their help is significant in problem solving in cases of interpreting data and information coming from sensors, suggesting potential diagnosis to clinicians, identifying potential risks and detecting behaviors, as well as helping managers to achieve business goals by improving planning and programming processes. Therefore, expert systems are often a basis or a component of *decision-support systems*: such systems allow to empower the classical information systems in companies, by including several modules and components to capture, organize and manage data and information to support strategic decision-making processes in business contexts.

The KBS are usually classified according to two complementary categorizations: on the one hand, the *potential application* is considered, on the other hand the *problem-solving methodology*: in other terms, the knowledge representation and reasoning method. As regards the application, the KBS can be divided in *knowledge capturing and storing* systems, *knowledge deployment and sharing* systems and *knowledge processing* systems. This classification allows to specifically focus on the specific ability of the considered system, throughout the whole process of collecting, managing and elaborating data and information to get useful knowledge. According to the problem-solving methodology, the KBS can be classified in *rule-based*, *case-based* and *model-based* systems. Such classification directly derives from the categorization of the reasoning methodologies. Indeed, the internal reasoning mechanisms exploited by the KBS can be based on logical rules, or on the use of *analogies* and *cases* to make inferences, or alternatively on the use of specific model-based descriptions. The following Figure is intended to help readers appreciate the different stages and concepts related to the knowledge representation and reasoning methodologies and systems (Fig. 2.2).

The use of such symbolic and knowledge management approaches is definitely useful for the representation and elaboration of knowledge in well-circumscribed and defined domain or contexts. The problems arise when it comes to learning from wide and complex sets of data, often heterogeneous in their structure or coming

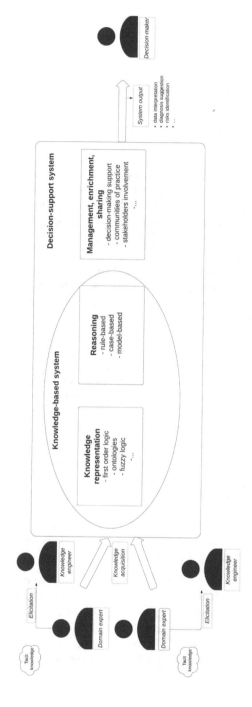

Fig. 2.2 Conceptual representation of the knowledge representation and reasoning process

from different sources. The ability to learn from past in data and to predict future behaviours, situations or states is typical of the other dominant paradigm of AI: connectionism, and, specifically, machine learning approaches.

2.7 Machine Learning and Deep Learning

In the current scenario of AI methodologies, approaches and algorithms, the lion's share of the dominant paradigm goes to machine learning. This is still true, but it should definitively not taken for granted, as AI is continuously undergoing a fast, dynamic and somehow unpredictable evolution, fostered by the advancements in technology, data sources and availability, organizations and society as a whole. Nevertheless, machine learning models are particularly relevant on the global scene of AI methods and algorithms thanks to their ability to analyze large sets of data, learn from them and predict new and unseen behaviors, or classify new instances in different environmental conditions.

In a general perspective, machine learning (ML) is the discipline devoted to designing, building and evaluating AI-based models which are able to learn from sample, training, pre-defined specific data, by identifying relevant *patterns* and behaviors, so as to make inferences from them and provide human decision-makers with significant automated classifications, suggestions or predictions. *Data mining* is the whole set of methods, algorithms and evaluation techniques to extract value and in insights from data. ML originates from the disciplinary areas of *statistical learning*, *computational learning* and *pattern recognition*: in this sense, it is inherently based on the use of statistical approaches to analyze sample data, perform relevant tests and experiments and draw problem-related inferences. Then, the whole design and programming procedures for such AI approaches are aimed at obtaining data-driven output or predictions, thus dynamically modifiable according to the evolution in the available datasets. The instances in the data are generally structured around specific *features* to represent them, which can be *categorical* or *numerical*, binary or continuous. The applications of ML methods are immense, ranging from speech recognition to computer vision, from strategic and operative business intelligence to fraud detection, from astrophysics analysis of the galaxies to financial applications.

Overall, ML methods can be classified in three *learning paradigms*, widely exploited in the literature and in the daily practice:

1. *Supervised learning* methods: this kind of approaches present pre-defined and already provided input and output variables, as sample already classified or analyzed instances, and the algorithm is able to generalize, by detecting the mapping function from input to output.
2. *Unsupervised learning* methods: in this case, the provided data are not labelled yet, thus the model is not provided with sample pre-analyzed data and results during the training process.

3. *Reinforcement learning* methods: such approaches are trained by providing them specific feedback and the intelligent system or agent strives to maximize some cumulative reward function to progressively optimize its output.

Therefore, while the working mechanisms of ML methods can significantly vary in a widespread range of possible algorithms, it is definitively clear that such AI approaches are able to *generate knowledge from data*, in a totally different way from knowledge-based systems and knowledge representation methods: this is the process of *knowledge discovery*, which helps practitioners, decision-makers and organization stakeholders to get significant, accurate and useful knowledge from data, with the aim to comprehend domain-related and environmental phenomena or behaviors, as well as to improve the quality and the effectiveness of decision-making processes. In a general perspective, ML is aimed at producing better outcomes for specified learning problems, by learning from past experience, in order to help humans to deal with large sets of data which require effective and efficient automated analysis in heterogeneous conditions, as in cases of uncertainty and complexity.

Supervised learning is widely diffused and exploited in the literature and in the practice. Basically, it consists of providing already labelled and categorized data and related instances, comprehending both their input and the desired output. Thus, the dataset is directly and naturally split into two parts, required for learning: the *training set* and the *test set*. The former is exploited to train the model by the means of the labelled *training examples*, while the latter is subsequently needed to allow the model to make new inferences on unseen data. In the ideal general process, the training of the model should provide the *mapping function* linking the input data and the desired output by examining the already labelled data, while the testing process should apply it to new instances to get predictions or insights. The SL models allow to deal with two macro-categories of problems: *classification* and *regression*.

Classification is the problem of categorizing sets of data into specific pre-defined classes: if there are only two classes, the problem is reduced to binary classification, which can be further extended to multi-class problems. Among the classification methods and algorithms there are non-probabilistic approaches like *support-vector machines*, but also and especially inherently probabilistic methods like *Naïve Bayes* classifiers, as well as *neural networks, rule-based* and *logic-based* approaches, or *decision trees*. Instead, regression is the problem of predicting real-valued *target response* to *predictor variables*, so as weighted sum of input features. The main aim is to detect and model the dependence of a regression target on some pre-defined considered features. While linear regression is endowed with relevant advantages thanks to its linearity and the consequent *transparency* of the linear effects of the different features, at the same time it requires handcrafted modifications to deal with non-linearities and the interpretation of the exploited weights is not necessarily intuitive. Thus, extended approaches like *generalized linear models* and *generalized addictive models* are often studied and exploited in the literature and in the daily practice.

Unsupervised learning is the learning paradigm for ML models which does not require already pre-labelled data or instances. In this case, the training set and the test set are present too, and the learning process is still based on the training of the algorithm on past experience and examples. The model is trained to discover by itself the relevant patterns and previously undetected insights related to the information extracted by data. Such learning paradigm is mainly exploited for dealing with the problems of *clustering*, to create groups of data which share similar characteristics, *dimensionality reduction*, to find the main factors of variability of data, and *anomaly detection*, to identify dissimilar examples among groups of instances. The most popular and well-known problem here is clustering: the issues related to grouping data by some specific characteristics to separate them are widely diffused in the practice and applications. In the K-means clustering approach, the objective is to create K-distinct groups, each one with a specific centroid, one for each cluster, and grouping data by minimizing the square error function. Instead, the problem of dimensionality reduction is generally addressed by themeans of approaches like the principal component analysis, helping to reduce two-dimensional into one-dimensional data, or neural networks like autoencoders, which are also useful to deal with anomaly detections and in many applications of the unsupervised learning paradigm.

Finally, *deep learning* is a part of ML approaches, including all the models based on *representation learning* and dealing with multiple *representation layers* to learn from data in complex situations, where multiple levels of abstractions and complexity are involved. In this situation, conventional, shallow machine learning models are inherently limited in their power and capacity. Therefore, deep learning approaches are useful to extract features implicitly, passing through a set of *hidden* layers in the structure of the network or algorithm involved, where the data are passed, elaborated and transformed. The *deep neural networks* are indeed extended neural networks with many input and output layers and data are transmitted and managed without loops in the learning process, in *feedforward* networks. Among the mostly diffused and exploited approaches there are *convolutional neural networks, recurrent neural networks* and *deep belief networks*. Notwithstanding the powerful ability of deep learning methods to analyze big data and make inferences in conditions of complexity and uncertainty, ML and also deep learning approaches are still limited and they suffer from several different problems, e.g. issues in *transparency, robustness, trustworthiness* and *reliability*. Therefore, after this general overview of both symbolical and sub-symbolical AI methods, it is now time to go further and explore the promises and challenges of neuro-symbolical approaches.

2.8   Neural Symbolic Computation

The goal of *neural symbolic computation* is to effectively integrate and exploit, at the same time, symbolic, logic and formal approaches to computation, together with connectionist, sub-symbolic, neural methods for enhancing the system power, cognitive abilities and accuracy in the output. Such methods strive to address the issues arising with both the first and the second wave of AI, to recall DARPA classification. In general, they try to meet the need for the integration of *knowledge representation* and *reasoning* mechanisms with deep learning-based methods. The aim is to provide researchers and practitioners with more accurate and also *explainable* systems, to enhance both the *predictive power* and the *interpretability* and the *accountability* of the systems, thus favoring the *trust* in the AI decision-making assistants for humans. While we will focus on the *explainability* issues in the next sections, here we just aim to provide a very short and general overview of the main pillars of such methodologies.

Neural-symbolic computing has the goal of integrating the two complementary cognitive abilities we already examined for symbolic and sub-symbolic approaches, respectively: *learning*, from the outer environment or the past experience, and *reasoning* from the result of the learning processes. Neural-symbolic computing tries to exploit both the advantages of learning in advanced up-to-date neural networks and reasoning in symbolic representation, leading to *interpretable* models. Therefore, the main idea behind neural-symbolic computation is the reconciliation of the symbolic and connectionist paradigms of AI under a comprehensive general framework. In this perspective, knowledge is represented by the means of symbolic formal mechanisms, e.g. *first order logic* or *propositional logic*, while learning and reasoning are computed by the means of adapted neural networks. Thus, the framework of neural-symbolic computation allows to effectively combine robust and dynamic learning with inference in neural networks. This increases both the computational capabilities and efficiency of the AI systems, the *robustness* of the algorithm and the accuracy of the output results, not least along with interpretability provided by symbolic knowledge extraction and reasoning by the means of formal theories and logic systems.

In a general perspective, neural-symbolical AI systems are able to deal with a broader spectrum of cognitive abilities than the previously presented methods, by exploiting the synergies between the capabilities of connectionist and symbolistic approaches. Moving step forward, the main characteristics of such systems are related to knowledge extraction and representation, reasoning, learning, with their applications and effects in different domains. Within the growing scenario of neural-symbolical computing, the *knowledge-based artificial neural network* (KBANN) and the *connectionist inductive learning and logic programming* (CIL^{2}P) systems are some of the most diffused models. KBANN is a system for *insertion, refinement* and extraction of rules from neural networks. The results of the design and development of KBANN suggested that neural-symbolical approaches can effectively improve the output of a learning system by the integration of *background knowledge* and learning from examples. The KBANN system was the first to set this research path, with significant applications in bioinformatics. KBANN also served as one of the inspirations for the design and construction of CIL^{2}P. This method integrates inductive learning from examples and background knowledge with deductive learning from *logic programming*. Basically, logic programming allows to represent a program by the means of formal logic. Thus, CIL^{2}P allows to represent background knowledge by a propositional general logic program, enables training from examples to refine the background knowledge and the subsequent knowledge extraction steps are performed by the use of a network, in a logic program.

One of the main challenges in neural-symbolic computation is the solution of the so-called *symbol grounding problem* (SGP). Essentially, this is the problem of endowing AI systems with the capacity to autonomously and automatically create *internal representations* to connect their manipulated symbols to their corresponding elements or *objects* in the external world. The methodology to ground symbols to environmental objects is a delicate and fundamental step in the design and development of a neural-symbolic AI system, often solved by grounding symbols to *real functions* or, in general, to some sort of *functionals* to encode the real-world elements characteristics and features. Solving the SGP enables effective propagation of information via symbol grounding, and, ultimately, the whole working mechanisms of the neural-symbolical network to capture and elaborate knowledge within the system. The way neural-symbolic approaches address and solve the SGP are among the differences among methods like *Deep Problog* and *logic tensor networks*.

The applications of neural-symbolic systems are in a wide range of fields, e.g. data science, learning in ontologies, training and assessment in simulators, and cognitive learning and reasoning methods for different applications. One of the most significant advantages of such systems are related to explainability. Differently from deep learning and *black-box* models, such systems can be explainable, thus allowing human decision-maker to understand the *reasons* behind their provided output. Indeed, the growing complexity of AI systems requires methods that can be able to explain users their working mechanisms and decisions. Thus, it is necessary and unavoidable to get a general picture of the fundamental role of *explainable AI* in the current scenario and evolution of AI systems.

2.9 Explainable AI

The rise and diffusion of eXplainable AI (XAI) is a crucial step in the current and future evolution of AI systems. Along with the neural-symbolical approaches, such methods can significantly be grouped into the so-called *third wave* of AI, as defined by DARPA. Thus, XAI strives to realize the big challenge of *contextual adaptation*, i.e. the construction of progressively explanatory methods for classes of real-world phenomena. While the field of XAI is actually not new in the literature and in AI research, nowadays its importance is constantly growing, by its ability to address the need for *interpretable, transparent* and *accountable* systems. The main reason behind explainable models is the general *opacity* and lack of *interpretability* in machine learning and deep learning models: despite the computing power and the high performances achieved by AI, and specifically by machine learning and deep learning systems, it is hard work to get insights from their internal mechanisms when trying to understand why some outcomes came from.

In order to address this issue, XAI is aimed at creating a set of AI techniques able to make their own decision more transparent and interpretable, so as to open the black box to share with users the way outputs come from. The final goal is to create completely explainable models and algorithms maintaining high performance levels. In this way, XAI allows user to enhance their interaction with the AI systems, by increasing their *trust* in the system mechanisms, thanks to the comprehension of the specific reasons why some output or decision was made. *Explainability* is badly needed in a wide range of applicative domains and regulations are starting to arise, requiring it as an essential and unavoidable feature of AI systems.

While there is not a single, unified, pre-defined and widely accepted formal definition of *explanation*, some key concepts emerge in the literature. To strive to define what precisely an explanation is the role of the user or *observer* is essential: in this perspective, an explanation aims at making the relevant *details* of an object clear or easy to understand to some observer. So, an explanation is able to provide users with the significant and needed information to allow them understand how the model works or why it made a specific decision or provided a particular output or suggestion. Moving a step forward, it is worth to highlight that an explanation is an *interface* between a human decision-maker and an intelligent assistant, enabling the human agent to understand some key relevant *proxies* to get the reasons of the system output, as well as the system internal working mechanisms. This adds a new dimension to the concept of explanation, showing new characteristics an explanation should have. It should be an accurate proxy, i.e. the explanation must be based on the model's mechanisms and the input used. In this perspective, the challenge of generating explanations requires to guarantee that they have accurate proxies, which is not granted in the state-of-the-art models. Such proxies are intrinsically related to the *features* managed by the model: an explanation usually relates the feature values of an instance to its model prediction in a way that can be easily understood by human decision-makers, practitioners or organization stakeholders.

In the literature, there are several attempts to provide classifications, categorizations and characterizations of XAI systems. Explainable recommendations can either be *model-intrinsic* or *model-agnostic*: in the former case, the output model is intrinsically interpretable, meaning that the decision mechanism is completely transparent providing explainability; in the latter case, instead, the output model provides the so-called *post-hoc explanations*, without any modification of the model itself. These two approaches can be conceptually related to a cognitive psychological root: in this perspective, the model-intrinsic models would be somehow similar to the human mind rational decisions, while the model-agnostic ones would somehow resemble the intuitive ways of deciding, followed by some search of the explanations. This classification is also reflected in the categorization of interpretable AI models, Interpretable machine learning techniques can generally be grouped into two categories: intrinsic interpretability and post-hoc interpretability, where the difference is in the moment when the interpretability is obtained and included into the model. Intrinsic interpretability is obtained by designing self-explanatory models incorporating directly interpretability. Among these approaches there are *linear models, decision-trees, rule-based* systems etc. In contrast, the post-hoc methods are created by defining another model to include explanations for an existing system.

Another important classification, taken from the specific sub-field of *explainable recommender systems*, is useful for conceptually categorizing two main relevant dimensions in the world of XAI: the dichotomy between the *information source* or the *display style* perspective of the explanations and the explainable model perspective. The former is the way the explanations are provided to users, e.g. *textual explanations*, *visual explanations* etc., while the latter is the specific model exploited for designing and developing the considered explainable system. Thus, while the display style represents the *human–computer interaction* perspective, while the model itself is related to the machine learning dimension of XAI research. Overall, the role of the interaction between the user and the model is fundamental, when it comes to explanations: the growing diffusion of XAI and its applications in a wide range of application domains requires a high-quality interaction and the effectiveness of the information and knowledge sharing processes between the intelligent assistants and the human-decision makers. Therefore, the rise and the evolution of the promising discipline of human—AI interaction is fundamental for the next stages in the AI history.

2.10 Human—AI Interaction

The current scenario and evolution of AI, as briefly and generally presented throughout this chapter, has nowadays led to focus the researchers' and practitioners' attention towards the role and the quality of the interaction between the AI systems and the human agents. While AI can be certainly considered as a form of automation, it is certainly not a mere technological means: rather, it is something more, enabling a continuous, non-linear and somehow unpredictable dynamics of the whole society. In a broader perspective, AI is becoming a real agent, not just supporting humans but interacting and exchanging information and knowledge with them. It is then necessary to embrace a wider perspective, where humans and AI work together, in order to enhance and empower the decision-making processes and the whole environments where such actors live and act. This poses relevant and unavoidable ethical issues, due to the possible risks deriving from somehow "equating" humans and AI. Whereas the ethical concerns are fundamental and should be definitively faced and solved by the means of a pre-defined and widely accepted framework, this is not the aim of this chapter. To our scope, it is sufficient to understand that humans and AI work, reason and act together in advanced and diffused socio-technical systems: better, in *socio-AI systems* that we could not ignore or underestimate. Furthermore, such systems will increasingly stimulate a wide variety of *emergent collective behaviors*, in a complex system perspective.

In this context, human—AI interaction (HAII) is an emerging and challenging discipline, aimed at studying, designing and evaluating interactive AI-based systems, as well as the potential relationships, issues and problems throughout the interaction, collaboration and teaming between humans and AI-based systems. HAII applies the classical paradigms and frameworks derived from human—computer interaction (HCI) to the cases where the technological interactive system is based on AI. Thus, the concepts related to *user—centered design, usability* and *user experience* are widely retained and adapted. The role of the users is fundamental: humans should be in the loop when designing and improving the AI systems modelling. In the field of HAII several frameworks for evaluating the application, interaction and the consequences of the exploitation of AI-based systems in heterogeneous application domains: one of the most important frameworks is the Artificial Intelligence Impact Assessment, which helps to map the potential benefits and threats of an AI application in a real-world situation, identify and assess the reliability, safety and transparency of AI systems and limit the potential risks in their deployment. Other frameworks and modelling instruments are aimed at identifying the current level of evolution and automation of the AI systems and functionalities, with respect to the role of the humans in the interaction: to this aim, the Parasuraman's model for types and levels of human interaction with automation can be applied to the assessment of the degree of automation of AI-based systems.

In this perspective, AI is increasingly becoming a *teammate* and a *decision-maker*, not just a mere *tool* or *assistant* or agent. To be able to design and develop reliable and trustworthy AI systems, as well as to enhance the kinds of interactions with humans, *guidelines* for HAII have been designed and applied. Such guidelines are meant to improve *fluency*, *effectiveness* and the overall *coordination* between humans and AI-based systems. They consider and strive to improve all the stages of the operations, before, during and at the end of the interaction and of the whole knowledge extracting and decision-making process. The main focus, again, is to put the user in the loop, meaning making clear the actions, decisions, contextually relevant information throughout all the process. Moreover, they strive to ensure that all the needed information and knowledge collection stages are well performed and that knowledge is accurately elicited, collected and transmitted: thus, they highlight the importance of learning from past feedback and from user behavior, as well as to remember the past interactions. The idea is to enable the AI system to adapt and evolve in its behavior, on the basis of the previous knowledge and experience accumulated, together with the human agent involved. Following such approach, there are several methodologies and frameworks to design the interaction between human actors and AI-based systems, e.g. smart and *conversational interfaces*, *anthropomorphism* and the application of *ethopoeia*, namely the attribution of human attitudes or intentions to machines, in the conceptual paradigm of the "*computer as social actors*" theory.

At the end of this journey, what's next? Well, there is much more waiting for us at the horizon! AI is currently shifting its role and characteristics in the whole society: starting from mere and inherently limited automated decision-support systems and technological infrastructure to interacting assistants, teammates and even *managers* and decision-makers. Thus, human and machine intelligences are going to combine in unpredictable and synergistic ways, reaching the so-called *hybrid intelligence*, where the human and the machine cognitive abilities are exploited together, in an unified and empowered environment, so as to achieve complex goals and reach superior results, while enabling continuous learning by a fluid and reciprocal knowledge and experience sharing between human and machine agents. And next? The rise of *collective hybrid intelligence* is going to revolutionize probably every aspect of human society, leading to an emerging, comprehensive and socio-AI systemic intelligence. The digital, social and cultural transformation stimulated by the diffused application of AI to human society arises questions that we need to answer, now and every day. Let us work and improve every day our comprehension and awareness of the potentialities and of the challenges of AI, to ensure that its power and advantages can serve to our sons and to all the new and future generations. This should always be our ultimate goal, as researchers, scholars and practitioners. As humans.

References

M.S. Abdullah, C. Kimble, I. Benest, R. Paige, Knowledge-based systems: a re-evaluation. J. Knowl. Manag. **10**(3), 127–142 (2006). https://doi.org/10.1108/13673270610670902

A. Agrawal, J. Gans, A. Goldfarb, in *Prediction machines: the simple economics of artificial intelligence* (Harvard Business Press, 2018)

R. Akerkar, P. Sajja, *Knowledge-Based Systems* (Jones & Bartlett Publishers Inc., USA, 2009)

R.J. Amineh, H.D. Asl, Review of constructivism and social constructivism. J. Soc. Sci. Literat. Languages **1**(1), 9–16 (2015)

E. Brynjolfsson, A.N.D.R.E.W. Mcafee, Artificial intelligence, for real. Harv. Bus. Rev. **1**, 1–31 (2017)

K. Butler, Connectionism, classical cognitivism and the relation between cognitive and implementational levels of analysis. Philos. Psychol. **6**(3), 321–333 (1993). https://doi.org/10.1080/095150 89308573095

D. Fasko Jr., An analysis of multiple intelligences theory and its use with the gifted and talented. Roeper Rev. **23**(3), 126–130 (2001)

N. Fleming, How artificial intelligence is changing drug discovery. Nature **557**(7706), S55–S55 (2018)

J.A. Fodor, Precis of the modularity of mind. Behav. Brain Sci. **8**(1), 1–5 (1985)

C.B. Frey, M.A. Osborne, The future of employment: how susceptible are jobs to computerisation? Technol. Forecast. Soc. Chang. **114**, 254–280 (2017)

A. Garnham, Cognitivism. in *The Routledge Companion to Philosophy of Psychology* (Routledge, 2019), pp. 99–110

Gartner, Gartner top 10 strategic technology trends for 2020. Gartner (2021). https://www.gartner. com/en/information-technology/insights/top-technology-trends. Last Accessed 14 Aug 2022

L. Gunderson, L.S. Siegel, The evils of the use of IQ tests to define learning disabilities in first-and second-language learners. Read. Teach. **55**(1), 48–55 (2001)

N. Haefner, J. Wincent, V. Parida, O. Gassmann, Artificial intelligence and innovation management: a review, framework, and research agenda☆. Technol. Forecast. Soc. Chang. **162**, 120392 (2021)

F. Van Harmelen, V. Lifschitz, B. Porter (Eds.), in *Handbook of knowledge representation*, vol 2 (Elsevier, 2008), pp. 1–1006

IDC, Worldwide semiannual artificial intelligence tracker (International Data Corporation, 2022)

L. James, L. James, S. Stuart, N. Roger, L.H. Christopher, Artificial intelligence: a general survey (Science Research Council, 1973)

S. Ji, S. Pan, E. Cambria, P. Marttinen, S.Y. Philip, A survey on knowledge graphs: representation, acquisition, and applications. IEEE Trans. Neural Netw. Learn. Syst. **33**(2), 494–514 (2021)

C.G. Langton, Studying artificial life with cellular automata. Physica D **22**(1–3), 120–149 (1986)

Weng et al., Autonomous mental development by robots and animals. Science **291**, 599–600 (2001)

J. Launchbury, A DARPA perspective on artificial intelligence. Technica Curiosa (2017). Online URL: https://machinelearning.technicacuriosa.com/2017/03/19/a-darpa-perspective-on-artificial-intelligence/. Last Accessed 14 Aug 2022

J. Lee, T. Suh, D. Roy, M. Baucus, Emerging technology and business model innovation: the case of artificial intelligence. J. Open Innov.: Technol. Market and Complexity **5**(3), 44 (2019)

Y.G. Lyu, Artificial intelligence: enabling technology to empower society. Engineering **6**(3), 205–206 (2020)

B. Marr, *Artificial Intelligence in Practice: How 50 Successful Companies Used AI and Machine Learning to Solve Problems* (Wiley, 2019)

J. McCarthy, M.L. Minsky, N. Rochester, C.E. Shannon, A proposal for the dartmouth summer research project on artificial intelligence, august 31, 1955. AI Mag. **27**(4), 12–12 (2006)

P. McLeod, K. Plunkett, E.T. Rolls, *Introduction to Connectionist Modelling of Cognitive Processes* (Oxford University Press, 1998)

A. Newell, J.C. Shaw, H.A. Simon, Report on a general problem sol[ving program. in *IFIP Congress*, June, vol 256 (1959), pp. 64

F. Rosenblatt, The perceptron: a probabilistic model for information storage and organization in the brain. Psychol. Rev. **65**(6), 386 (1958)

P. Salovey, J.D. Mayer, Emotional intelligence. Imagin. Cogn. Pers. **9**(3), 185–211 (1990)

M. San Miguel, J.H. Johnson, J. Kertesz, K. Kaski, A. Díaz-Guilera, R.S. MacKay, D. Helbing, Challenges in complex systems science. The Europ. Phys. J. Special Topics **214**(1), 245–271 (2012)

K. Schwab, in *The fourth industrial revolution*. Currency (2017)

S. Shanmuganathan, Artificial neural network modelling: an introduction. in *Artificial Neural Network Modelling. Studies in Computational Intelligence*, ed. by S. Shanmuganathan, S. Samarasinghe, vol 628. (Springer, Cham, 2016). https://doi.org/10.1007/978-3-319-28495-8_1

Statista, Revenues from the artificial intelligence (AI) market worldwide from 2018 to 2025. Statista (2022)

G. Von Krogh, Artificial intelligence in organizations: new opportunities for phenomenon-based theorizing. Acad. Managem. Discoveries **4**, 404–409 (2018)

H. Wang, B. Raj, On the origin of deep learning (2017). arXiv preprint arXiv:1702.07800

J. Weizenbaum, ELIZA—a computer program for the study of natural language communication between man and machine. Commun. ACM **9**(1), 36–45 (1966)

K. Westcott Grant, Netflix's data-driven strategy strengthens claim for 'best original content' In 2018. Forbes (2018). https://www.forbes.com/sites/kristinwestcottgrant/2018/05/28/netflixs-data-driven-strategy-strengthens-lead-for-best-original-content-in-2018/?sh=8f21fc13a94e. Last Accessed 14 Aug 2022

World Economic Forum, The future of jobs 2017. (World Economic Forum, 2017)

J. Zhang, B. Chen, L. Zhang, X. Ke, H. Ding, Neural, symbolic and neural-symbolic reasoning on knowledge graphs. AI Open **2**, 14–35 (2021)

M. Ziegler, E. Danay, M. Heene, J. Asendorpf, M. Bühner, Openness, fluid intelligence, and crystallized intelligence: toward an integrative model. J. Res. Pers. **46**(2), 173–183 (2012)

Luca Marconi is currently PhD Candidate in Computer Science at the University of Milano-Bicocca, in the research areas of Artificial Intelligence and Decision Systems, as well as Business Strategist and AI Researcher for an AI company in Milan. He also gained experience as Research Advisor and Consultant for a well-known media intelligence and financial communication private consultant in Milan. He holds a master of science and a bachelor of science degrees in Management Engineering, from the Polytechnic University of Milan, and a master of science degree in Physics of Complex Systems, from the Institute of Cross-Disciplinary Physics and Complex Systems of the University of the Balearic Islands (IFISC UIB-CSIC). He also holds a postgraduate research diploma from the Advanced School in Artificial Intelligence, Brain, Mind and Society, organized by the Institute of Cognitive Sciences and Technologies of the Italian National Research Council (ISTC-CNR), where he developed a research project in the Computational Social Science area, in collaboration with the Laboratory of Agent Based Social Simulation of the ISTC-CNR. His research interests are in the fields of artificial intelligence, cognitive science, social systems dynamics, complex systems and management science.

Chapter 3
Financial Markets: Values, Dynamics, Problems

Juliana Bernhofer⊙ and **Anna Alexander Vincenzo**⊙

Abstract The 2020 global stock market crash triggered by the coronavirus outbreak has led to an intertwined and still unfolding series of social, economic, and financial consequences. We will review several aspects of the recent pandemic crisis and critically compare them to extant findings on the effects of financial breaks, fear, and turbulences (global financial crises, terrorism, and previous pandemics) on financial markets. With the increase of uncertainty and volatility, the Covid-19 pandemic also affected investors' risk attitudes, trust, and confidence. We will provide insight into these behavioral aspects and how they ultimately affect financial decision-making. Moreover, a unique side effect of the government restrictions is the unprecedented increase of digitalization: never before have global citizen of all ages been cornered into using online services and digital tools. We will discuss the implications of this new phenomenon for the financial market, its effect on social inequality, and the scenario of opportunities it entails.

Keywords Financial markets · Risk · Trust · Resilience · Digital literacy

3.1 Introduction

In late 2019, Covid-19 had started to spread. Due to severe pneumonia, the virus led to an extraordinary rate of death between the elderly and the most vulnerable in society, leading to an international social distancing and home isolation of more than 2 billion people worldwide, with tremendous social, political, and economic

J. Bernhofer (✉)
Department of Economics, University of Modena and Reggio Emilia, "Marco Biagi", Via Jacopo Berengario, 51, 41121 Modena, Italy
e-mail: juliana.bernhofer@unive.it

Department of Economics, Ca' Foscari University of Venice, Cannaregio 873, 30121 Venezia, Italy

A. Alexander Vincenzo
Università degli Studi di Padova, Via del Santo 33, 35123 Padova, Italy
e-mail: anna.alexandervincenzo@unipd.it

© The Author(s), under exclusive license to Springer Nature Switzerland AG 2023
F. Cecconi (ed.), *AI in the Financial Markets*, Computational Social Sciences,
https://doi.org/10.1007/978-3-031-26518-1_3

consequences (Goodell 2020; Sharif et al. 2020). Moreover, one of the commercial effects of this virus was an unexpected disruption in the flow of goods and services, commodity prices and financial conditions, creating economic disasters across many countries because of its impact on the production and supply chains in China, the United States, Europe, and Japan, and other important economies in the world (IMF 2020).

The outbreak of Covid-19 was a shock to market participants (Ramelli and Wagner 2020) and the anticipated real effects of the health crisis were amplified through financial channels. In contrast to the Global Financial Crisis of 2007, Covid-19 primarily was a shock to the real economy, initially unrelated to the availability of external financial means though the market's reaction anticipated a potential amplification of the real shock through financial restrictions.

Analyzing the impact of this unanticipated rare disaster on financial markets, we can outline three phases in the progression of the pandemic.[1] On December 31, 2019, cases of pneumonia detected in Wuhan, China, were first reported to the World Health Organization (WHO), and on January 1, 2020, Chinese health authorities closed the Huanan Seafood Wholesale Market after it was discovered that the wild animals sold there may be the source of the virus (World Health Organization 2020a). The first trading day was January 2, hence the starting point of the first period.

Second, on January 20, Chinese health authorities confirmed human-to-human transmission of the coronavirus, and the following day the WHO issued the first situation report on the outbreak (China's National Health Commission 2020; World Health Organization 2020b). Monday, January 20 was a national holiday in the United States; therefore the following day marked the beginning of the second phase.

Third, on Sunday, February 23, Italy placed almost 50,000 people under strict lockdown in an attempt to control the outbreak after registering its first deaths from coronavirus on Saturday, February 22 (Italy's Council of Ministers 2020). February 24 was the first trading day in this phase and the situation plummeted in the following weeks. On Monday, March 9 the Dow Jones Industrial Average (DJIA) fell 2014 points, a 7.79% drop. Two days later, on March 11, the World Health Organization classified Covid-19 as a pandemic. On March 12, 2020, the DJIA fell 2352 points to close at 21,200. It was a 9.99% drop and the sixth-worst percentage decrease in history. Finally, on March 16, the DJIA collapsed by nearly 3000 points to close at 20,188, losing 12.9%. The fall in stock prices forced the New York Stock Exchange to suspend trading several times during those days.

In March 2020, the world witnessed one of the most dramatic stock market crashes so far. In barely four trading days, the DJIA plunged 6400 points, an equivalent of roughly 26%. The sharp fall in stock market prices was caused by the authorities' decision to impose strict quarantines on the population and ordered a lockdown of most business activities to limit the spread of the Covid-19 virus. Lockdowns imposed

---

[1] Ramelli and Wagner 2020 describe the evolution of the capital market reaction to the outbreak, recognizing three phases: Incubation (Thursday January 2 to Friday, January 17), Outbreak (Monday, January 20 to Friday, February 21), and Fever (Monday, February 24 through Friday, March 20).

Fig. 3.1 The Dow jones industrial average (DJIA) during the first quarter of 2020

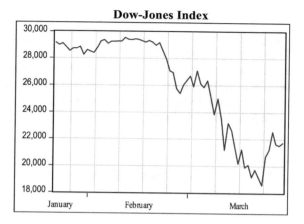

to business activities put on hold most of the business operations, and companies reacted by adjusting their labor costs to compensate the revenue shock. The layoff of employees led to a sharp reduction in consumption and economic output, lowering companies' stream of expected future cash flows.

The lowest point was hit by the DJIA on March 20 at the end of the trading week. The following Monday, March 23, the Federal Reserve Board announced major interventions in the corporate bond market and the index began to rise again. The evolution of the index during the first quarter of 2020 is shown in Fig. 3.1.

However, the Covid-19 pandemic did not affect all industries equally[2] and the lockdowns did not result in negative price changes across all industries. Some sectors, in fact, have benefitted from the pandemic and the resulting lockdowns. Mazur et al. 2021 focused on S&P 1500 and found that firms that operate in crude petroleum sector were hit hardest and lost over 60% of their market values in a day. In contrast, firms in natural gas and chemicals sectors improved their market valuations and earned positive returns of more than 10%, on average. Further, they study industry-level patterns and show that during March 2020 stock market crash, the best-performing industries include healthcare, food, software, and technology, as well as natural gas, which yield a positive monthly return of over 20%. On the other hand, sectors including crude petroleum, real estate, hospitality, and entertainment plunged over 70%.

If it's unsurprising that Covid-19 captured the attention of the financial markets, it is interesting to note when market participants became more preoccupied with the development of the situation. One measure used in research to gauge retail investor interest in a topic was developed by Da et al. 2015 and it uses the intensity of search on Google. Ramelli and Wagner 2020 exploit Google Trends search to capture investor sentiment and show that Global Google search attention on coronavirus spiked, especially after March 9, while the number of conference calls of international firms

[2] https://www.mckinsey.com/business-functions/strategy-and-corporate-finance/our-insights/the-impact-of-covid-19-on-capital-markets-one-year-in.

and the number of conference calls covering the coronavirus topic also increased, but spiked earlier on at the end of February. In their study, they find that initially, as China was effectively shut down, investors avoided U.S. stocks with China exposure and internationally oriented companies; as the virus situation in China improved relative to the situation in Europe and in the United States, investors perceived those companies more favorably again. As the virus spread in Europe and the United States, leading to lockdowns in these economies, markets moved feverishly.

3.2 How Did Financial Markets React to Previous Pandemics?

Few studies analyze the impact of epidemics or pandemics on the financial markets. Recent studies put the coronavirus pandemic into perspective and analyzed its consequences compared to previous pandemics of the twentieth century.

The impact of an infectious disease outbreak is characterized by high uncertainty, more so when the outbreak involves a new disease with an unquantified mode and rate of transmission, infectivity, and lethality (Smith 2006). According to (Baker et al. 2020), the impact of Covid-19 pandemic outbreak on the stock market is unprecedented.

In Fig. 3.2, we show the evolution over time of S&P 500 actual real prices. Stock-market returns for 1918 are hard to obtain and would have been affected by the end of the First World War, which took place between 1914 and 1918. The S&P 500 Index can be tracked back to 1871; it fell by 24.7% in 1918 and rose by 8.9% in 1919.

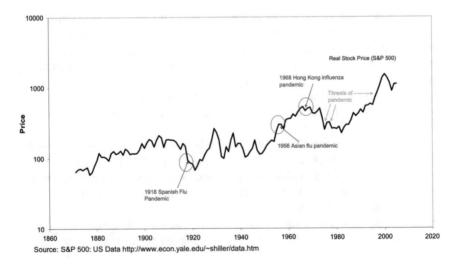

Fig. 3.2 S&P500 history 1860–2020

The 1957 Asian flu pandemic was the second most significant outbreak in recent history with approximately 1–2 million victims worldwide. This pandemic was first identified in the Far East in February 1957. After an initial bear market phase, the S&P 500 Index fully recovered by the last quarter of 1958.

In early 1968, the Hong Kong influenza pandemic was first detected in Hong Kong. Deaths from this virus peaked in December 1968 and January 1969. The S&P 500 Index rose by 12.5% in 1968 and by 7.4% in 1969. In 2003—the year that the SARS epidemic was first reported—the MSCI Pacific ex Japan Index fell by 12.8% from 14 January to 13 March. However, markets subsequently rallied strongly and, for the year as a whole, the index returned 42.5%.

Another study by Correia et al. 2022 looks at the impact of government interventions such as social distancing, aimed at reducing mortality, on the economic severity of the spread of a pandemic. The focus of their research is the 1918 Flu Pandemic that killed 675,000 people in the U.S., or about 0.66% of the population. Most deaths occurred during the second wave in the fall of 1918. In response, major U.S. cities implemented a range of actions to contain the spread of the virus, with different speed and stringency. These included school, theater, and church closures, public gathering bans, quarantine of suspected cases, and restricted business hours. Their results show no evidence that cities that enacted more stringent containment measures performed worse in the years after the pandemic. In fact, their findings suggest that these containment measures reduced the transmission of the virus without intensifying the pandemic-induced economic downturn.

In the aftermath of the SARS outbreak, studies like (Beutels et al. 2009; Sadique et al. 2007; Smith 2006) highlighted the economic implications of pandemics. Outbreaks like SARS are predicted to generate immediate direct economic effects from the impact of the disease itself (health sector costs related to treatment of those infected), from policies to contain disease (public sector and private costs related to quarantine, isolation and school closure) and from the fear that is generated by the outbreak, which will affect individual behavior. The presence of a health threat can influence the general expectations and behavior of consumers and investors and, therefore, have an impact far beyond the direct reduction in productivity from sick patients (Sadique et al. 2007). For instance, for SARS, it has been observed that by far the largest economic impact occurred by reduced local and international travel and reduced engagement in non-vital consumption activities such as restaurants and bars, particularly in May 2003, as well as postponement of consumption of mostly durable goods (Richard et al. 2008). Beutels et al. 2009 underline the importance of acknowledging consumers and investors adaptation to public health emergencies to provide useful economic analysis in the context of pandemics.

A decade before the Covid-19 outbreak, (Keogh-Brown et al. 2010) studied the possible economic impact on the UK economy of potential contemporary pandemics. They based their estimates using epidemiological data from past pandemics, such as the 1957[3] and 1968[4] influenza pandemics and applied a quarterly macroeconomic model of the UK economy. They used the latter two pandemics of the twentieth century as a base case to construct further disease scenarios by increasing the severity of the disease parameters, thus allowing to investigate what the economic effect of a more severe influenza pandemic might be.

The choice in their research to use the 1957 and 1968 influenza pandemics, in place of the 1918 pandemic, is driven by the observation that given the modern changes in societal behavior and public health policy, the economic impact of a recurrence of the Spanish flu would be very different in the current economy. However, by using the more recent pandemics as a base case for disease scenarios and extrapolating beyond the severity of more serious pandemics, such as 1918, they are able to generate some useful hypotheses and stimulate thinking on the potential economic impact of infectious disease outbreaks (Keogh-Brown et al. 2010).

Keogh-Brown et al.'s model provides different outputs according to the severity of each scenario in terms of clinical attack rate and case fatality rate. The shock to the working population is expressed as deaths, direct absenteeism (caused by infection) and indirect absenteeism (caused by school closure and 'prophylactic' avoidance of work). The results of the model, of course, vary according to the different values attributed to the severity of the outbreak, but provide interesting conclusions in terms of GDP loss, consumption decline, and inflation. Most importantly, these results clearly indicate the significance of behavioral change over disease parameters.

3.3   On Natural Disasters and Terrorist Attacks

While there is limited prior literature on how epidemics, let alone pandemics, impact financial markets, imperfect parallels can be drawn from other forms of natural disasters (Goodell 2020). Markets react to natural disasters such as earthquakes and volcanos, as well as air disasters, and more recently acts of terrorism.

Niederhoffer 1971 was the first to analyze the price impact of various crises, studying events ranging from Kennedy's assassination to the beginning of the Korean War. Niederhoffer found that these world events have a discernible influence on the movement of the stock market averages. Barrett et al. 1987 examined a sample of 78 fatal commercial airline crashes that occurred in the period from 1962 through 1985 and found that the immediate negative price reaction to airline crashes is significant for only one full trading day after the event occurs. They also analyzed market responses after the initial reaction period and found no evidence of under-reaction or overreaction in the initial response period, which is in line with the idea of immediate

---

[3] In February 1957, a new influenza A (H2N2) virus emerged in East Asia, triggering a pandemic ("Asian Flu"). It was first reported in Singapore in February 1957, Hong Kong in April 1957, and in coastal cities in the United States in summer 1957. The estimated number of deaths was 1.1 million worldwide. https://www.cdc.gov/flu/pandemic-resources/1957-1958-pandemic.html.

[4] The 1968 pandemic was caused by an influenza A (H3N2) virus. The estimated number of deaths was 1 million worldwide. https://www.cdc.gov/flu/pandemic-resources/1968-pandemic.html.

price adjustment in the market. Shelor et al. 1992 examined the impact that the earthquake that struck California on October 17, 1989, had on real estate-related stock prices. Their findings indicate that the earthquake conveyed new information relevant to the financial market, which was manifested in statistically significant negative stock returns for firms that operated in the San Francisco area. Real estate-related firms operating in other areas of California were generally unaffected by the earthquake and did not experience any significant price reactions. Worthington and Valadkhani 2004 measured the impact of a wide variety of natural disasters, including severe storms, floods, cyclones, earthquakes, and wildfires, on the Australian capital markets. Their results suggest that wildfires, cyclones, and earthquakes have a major effect on market returns, whereas severe storms and floods do not.

The indirect economic consequences of terrorism on global financial markets have received considerable attention in the academic literature over the past few years. Several studies have examined the effects of terrorism on stock markets. Research on the impact of terrorist events on financial markets might provide some parallel, as terrorist events, while localized in their initial manifestation are by their nature designed to create a widespread change in public mood. Negative consequences can be observed on major macroeconomic variables with negative shocks to consumer and investor confidence and thereby also to the economic outlook and financial markets (Frey et al. 2007; Johnston and Nedelescu 2006).

A number of empirical studies have analyzed the stock market effects of terrorist attacks. Eldor and Melnick 2007 used daily data to analyze how the Tel Aviv Stock Exchange reacts to terror events. They focused on 639 attacks between 1990 and 2003 and found that suicide attacks had a permanent effect on the stock market. Karolyi and Martell 2006 looked at the stock price impact of 75 terrorist attacks between 1995 and 2002, focusing on publicly listed firms. Their cross-sectional analysis of the firm's abnormal stock returns suggests that the impact of terrorist attacks differs according to the home country of the target firm and the country in which the incident occurred. Attacks in countries that are wealthier and more democratic were associated with larger negative share price reactions. Furthermore, the results of their study suggested that human capital losses, such as the kidnapping of company executives, are associated with larger negative stock price reactions than physical losses, such as bombings of facilities or buildings.

Chen and Siems 2007 used event study methodology to assess the effects of terrorism on global capital markets. They examined the US stock market's response to 14 terrorist attacks dating back to 1915, and in addition analyzed the global capital market's response to two more recent events—Iraq's invasion of Kuwait in 1990 and the September 11, 2001 terrorist attacks. The results of that study suggest that the US capital markets are more resilient than in the past and tend to recover more quickly from terrorist attacks than other global capital markets. This latter finding is partially explained by a stable banking/financial sector that provides adequate liquidity to encourage market stability and minimize panic.

Moreover, Drakos 2004 and Carter and Simkins 2004 studied the specific impact of the September 11 attacks on airline stocks. The latter focused on the significant emotional impact of the attacks on investor psychology by testing whether the stock price reaction on September 17, the first trading day after the attack, was the same for each airline or whether the market distinguished among airlines based on firm characteristics. Their cross-sectional results suggest that the market was concerned about the likelihood of bankruptcy in the wake of the attacks and distinguished between airlines based on their ability to cover short-term obligations. Drakos focused on the risk characteristics of airline stocks and reported a terrorism-induced structural shift in financial risk in the aftermath of the September 11 terror attacks. Drakos reported that both the systematic and idiosyncratic risk of airline stocks increased substantially since September 11, 2001. Conditional systematic risk more than doubled, which could have implications for both firms and investors. Portfolio managers need to reconsider their allocation of money to airline stocks, and listed airlines will face higher costs when raising capital.

The impact of terrorism on financial market sentiment is also examined in Burch et al. 2003 and Glaser and Weber 2005. Burch et al. 2003 find a significant increase in closed-end mutual fund discounts in the aftermath of the September 11 attacks, and thereby conclude that these attacks caused a negative shift in investor sentiment. Glaser and Weber 2005 use questionnaire data to analyze the expectations of individual investors before and after the September 11 attacks. Somewhat surprisingly, they find that the stock return forecasts of individual investors are higher and the differences of opinion lower after the terrorist attacks. Glaser and Weber 2005, however, also report a pronounced increase in investors' volatility expectations.

Karolyi 2006 discusses the "spillover effects" of terrorist attacks and whether research on this topic suggests a broad-based or "systematic" contribution of potential terrorism to overall risk. His conclusion is that the evidence is quite limited, but there have been few tests that have examined volatility or beta risks with asset-pricing models. Choudhry 2005 investigated, post-September 11, a small number of US firms in a variety of different industries to see if this terrorist event affected a shift in market betas, with mixed findings. Hon et al. 2004 find that the September 11 terrorist attacks led to an increase in correlations amongst global markets with variations across global regions. A number of other papers present a mixed picture of how much terrorist acts have spilled over into changes in the nature of financial markets (Chesney et al. 2011; Choudhry 2005; Corbet et al. 2018; Nikkinen and Vähämaa 2010).

Some papers suggest that the downturn in markets with respect to terrorist events is rather mild. According to (Brounen and Derwall 2010), financial markets rebound within the first weeks of the aftermath of unanticipated events such as earthquakes and terror attacks. In their study, they compare price reactions of terror attacks to those of natural disasters and find that price declines following terror attacks are more pronounced. However, in both cases prices recover shortly after the event. They also compare price responses internationally and for separate industries and find that reactions are strongest for local markets and for industries that are directly affected by the attack. The September 11 attacks turn out to be the only event that caused long-term effects on financial markets, especially in terms of industry's systematic risk.

3.4 Risk, Trust, and Confidence in Times of Crises

How do risk preferences, attitudes toward trust and confidence in the government affect financial decision-making? In turn, to what extent do exogenous shocks such as the recent Covid-19 pandemic impact on these decisive parameters in the short and in the long run?

Exposure to disasters such as earthquakes, flooding, terrorist attacks and pandemics has an impact not only directly on the victims and their families, but on the whole society and aggregate preferences.

Risk preferences, Trust and Animal Spirits

Ahsan 2014 shows that anxiety, as a consequence of extreme events such as cyclones, increases people's risk aversion. Anxiety however affects stock market performance not only through the channel of risk preferences; post-traumatic stress and fear affect rationality and also introduce significant bias in interpreting reality. Engelhardt et al. 2020 analyze data from 64 countries covering 95% of the world's GDP to assess whether the stock market crash in 2020 was driven by rational expectations[5] or an increased attention to news. Similarly to (Ramelli and Wagner 2020), the latter was measured by assessing abnormal Google search volume within the assessed time frame. The authors find that one standard deviation increase in news attention leads to a 0.279 standard deviation decrease in market returns, whereas one standard deviation increase in the authors' rational expectations measure leads to a decrease of 0.131 standard deviations of market returns. The cost for the US stock market associated to the hype of negative news was estimated to be USD 3.5 trillion until April 2020. Also (Al-Awadhi et al. 2020) find negative effects on stock market returns in reaction to both total confirmed cases and daily Covid-19 deaths in the Chinese stock market and (Albulescu 2021) find similar results for the financial markets in the Unites States. Shear et al. 2020 compare different cultures and their level of uncertainty avoidance. The higher the tendency of trying to mitigate uncertain situations, the higher the search volume for additional information. The results are in line with previous reports: national culture, in association with the uncertainty avoidance index, is a significant predictor of stock market returns in times of crises, in that cultures who are more risk and uncertainty averse tend to seek for more information, ultimately increasing their level of anxiety which in turn has negative effects on stock market outcomes. Similarly, also (Bernhofer et al. 2021) show that speakers of languages defined by a more intensive use of non-indicative (*irrealis*) moods have a higher level of risk aversion and invest less in risky assets. This is due to the fact that, semantically, the unknown is more precisely defined, hence increasing the salience of their regions of uncertainty.

---

[5] Rational expectations were proxied with growth rates from the exponential fit and an epidemiological model for the spread of diseases (SIR model for susceptible-infectious-recovered). Even though this approach represents the best proxy for rational expectations, the authors underline that rationality here is, in fact, *bounded* as these models might turn out to be incorrect and lead to non-rational outcomes.

The idea that people are affected by their emotions when making financial decisions was first described by John Maynard Keynes in 1936 in his prominent publication *The General Theory of Employment, Interest and Money*. In what most likely represents the origin of behavioral finance, he illustrates how "animal spirits", or basic instincts and emotions, affect consumer confidence. In what was observed by Hippocrates from Kos in 400 B.C. and Claudius Galen from Pergamon in 200 A.D. as *pneuma psychikon* (or "animal spirits"), an invisible entity located within the cerebral ventricles and which moves through muscles and nerves (Herbowski 2013; Smith et al. 2012). This approximation however appears to be indeed the original idea of the biological explanation behind the psychological reaction to crises and uncertainty as a "spontaneous urge to action". One of the reactions to shocks hence appears to lie in the production of the "stress hormone", cortisol, which increases as a response to e.g. market volatility and has been shown to decrease risk propensity (Kandasamy et al. 2014). Also measured anxiety and its consequences on financial decision-making can be, at least in part, explained by the surge of cortisol levels as small probabilities tend to be overweighted as a response to increased levels of stress. The latter phenomenon is more commonly observed in men compared to women.

Moreover, the human tendency of paying attention to news instead of following data-driven models is what Christopher A. Sims in the early 2000s describes with his renowned theory of rational inattention which has since been discussed and applied by numerous researchers. According to Sims' thesis, investors do not have the information processing capacity in terms of data, time, and cognitive abilities to run highly complex models, so they take a shortcut, relying on media reports. While this is a money-saving technique on one hand, one of the major drawbacks of this approach is the exposure to *negativity bias* as the result of increased attention to pessimistic news. This, in turn, leads to the formation of pessimistic expectations and distortive effects in financial markets.

Trust and Confidence in the Government

Engelhardt et al. 2021 address the issue of behavioral aspects in financial decision-making from a slightly different angle by investigating a society's pre-existing inherent level of trust and its resilience against the effects of the pandemic. The authors analyze data from the World Value Survey covering 47 countries and the number of confirmed Covid-19 cases from the end of January to the end of July 2020. Trust was measured with two different indices, societal trust—or trust in other people—and confidence in the government. Market Volatility was calculated as the 5-day moving average volatility and Covid-19 cases were adjusted by population size. Stock markets were found to be significantly less affected by new Covid-19 case reports in those countries where societal trust and confidence in the government were higher. The authors hypothesize that the transmission mechanism passes through the level of uncertainty which is lowered by the confidence in the government's rules as well as the level of trust in fellow citizens obeying new rules and prescriptions. The role of an active government in building trust and mediating emotional responses, or "animal spirits", to crises was underlined by Akerlof and Shiller 2010. It should therefore act as a responsible parent to the economy, mitigating its reactions to critical events.

3.5 Digital Literacy and Covid-19

The Great Surge

Research on the Covid-19 pandemic usually comes with a bitter aftertaste: the death toll, the effects on mental health in all age groups, the loss of jobs and the closure of firms. Industries relying on face-to-face interaction suffered significant reductions in revenue, as opposed to realities more able to adapt to a remote market, such as Zoom Video Communications, Netflix, and Amazon. Miescu and Rossi 2021 conduct some impulse response function analyses and show that Covid-19-induced shocks reduced employment of poor households nearly twice as much as that of rich households. Private expenditure, on the other hand dropped by nearly 50% in rich households compared to poor households which is likely to be due to differences in the consumption portfolio with the entertainment and hospitality sector being temporarily unavailable.

As a consequence of social distancing, school closures and other restrictions, a renowned and widely discussed externality of the first lockdown was that businesses of all sizes and private citizen of all age groups were forced to navigate the online world in order to overcome the physical limitations. The adaptation to the digital environment happened nearly overnight and has acted as a catalyst (Amankwah-Amoah et al. 2021) for the use of a wide range of digital tools previously unknown or ignored. Governments and firms alike had to swiftly move from paper-based procedures to online services to reach their targets and private citizens of all age groups, social backgrounds and educational categories had to boost existing digital skills or acquire new knowledge to benefit from public aid, and maintain access to healthcare and participate in the market. There was not much space and time for resistance as the pandemic hit our entire social system.

In Fig. 3.3, we compare interviews conducted through the Survey of Health, Ageing and Retirement in Europe (Axel Börsch-Supan 2022). SHARE is a biennial cross-national panel database on socio-economic status, family networks, individual preferences and health of individuals that are 50 years and older, hence representing one of the more vulnerable age groups. The first round of questionnaires (i.e., Wave 1) was conducted in 2004 and involved 11 European countries and Israel. The fieldwork of the 8th wave was supposed to be finalized during spring 2020, but was interrupted by the pandemic in March 2020. In June of the same year, CATI[6] questionnaires were conducted with some respondents who had already been interviewed and some who had not answered yet. In the two graphs, we compare the pre-covid computer skill levels in Europe among the older share of the population with the post-covid self-elicited improvement of computer skills after the first lockdown 2020. By comparing countries with low pre-existing digital competencies to countries that entered the crisis with a high level of computer skills, we observe a tendency toward closing the digital gap as the former countries show a stronger increase in their skills on average.

---

[6] Computer-assisted telephone interview.

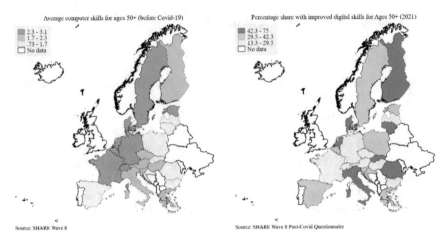

Fig. 3.3 Digital skills before and after the pandemic

A second econometric assessment presented in Table 3.1 sheds some light on the social aspect. We analyze the determinants of improved digital literacy after the Covid-19 pandemic, such as gender, literacy, household income level, and income inequality at the country level. On one hand, women benefit nearly twice as much as men and are also those who had a lower ex-ante digital literacy, but on the other hand, those who have better reading skills benefit slightly more from the digital transformation initiated by the pandemic. Interestingly, there does not appear to be any differential effect for respondents with a monthly income higher than the median of their country, but more importantly that countries characterized by higher levels of income inequality are on average 7.5 times more likely to have experienced an increase in digital literacy during the lockdown.

3.6 The Great Reset

Social sustainability is among the main goals of The Great Reset Initiative promoted by the World Economic Forum (WEF). The aim is to rise from the Covid-19 setback by creating a stakeholder economy that is more resilient, equitable, and sustainable. Growth should be greener, smarter, and fairer, in line with the United Nations 2030 Sustainable Development Goals.

A positive externality is provided by the digitalization process during the first lockdown in 2020: as we have seen earlier, who more frequently experienced increases in their digital skills during the lockdown are women and those who live in countries with a higher level of inequality. Also, those who have an educational level benefitted slightly more from the digitalization process, which means that skills are self-reinforcing. Knowledge makes it easier to acquire more knowledge.

Table 3.1 Some determinants of improved digital literacy

	(1)	(2)	(3)
Female	1.858***	1.781***	1.809***
	0.168	0.160	0.160
Literacy	–	1.118**	1.105**
	–	0.049	0.049
High Income	–	–	1.257
	–	–	0.186
High Gini	–	–	7.548***
	–	–	0.054
Constant	0.407***	0.207***	0.189***
	0.042	0.028	0.026
N. Observations	1992	1992	1992
N. Countries	24	24	24
Pseudo R2	0.016	0.055	0.057
Estimation method	Logit	Logit	Logit

Notes The dependent variable is "Has improved PC skills after Covid-19". The method of estimation is Logit with the coefficients reported as odds ratios. Robust standard errors are clustered at the country level. Country fixed effects have been included in all models. Literacy is measured on a scale from 0 (Fair) to 4 (Excellent). Reference categories for Female is Male. High Income $= 1$ if the respondent's income is higher than the country median. High Gini $= 1$ if income inequality is higher than the median of other countries in the sample

Significance levels: * $p < 0.1$, ** $p < 0.05$, *** $p < 0.01$

However, the role of the government is fundamental not only in creating trust, as described in the previous section, but also equally distributing and fostering digital skill acquisition, digital inclusion, and connectivity.

One of the drivers of wealth inequality is unequal access to traditional forms of credit. Small businesses and private citizens often cannot provide the sufficient proof of creditworthiness and without financial resources, a significant share of the population is cut out from the market thus hindering innovation and productivity. The most affected share of the population is represented by citizen with low income, women, and ethnic minorities. Artificial Intelligence has the potential to overcome these limitations as potential recipients can be evaluated in different domains: the creditworthiness of an individual is therefore not limited to their financial history but could extend, for example, to their social network activity and behavior, geographical location, and online purchases. Under the condition that equal access to digital technology is granted, the recent developments allow for a worldwide inclusion into financial, educational, and commercial realities of disadvantaged and geographically remote populations. In Bangladesh for example, iFarmer is a crowdfunding platform that puts together female cattle farmers targeting potential investors. Digital lending has experienced a significant increase with the onset of the pandemic with the special form of Fintech, P2P (peer-to-peer) lending over apps, becoming more frequently an alternative to banks or other financial intermediaries (Arninda and Prasetyani 2022). Loans in such platforms are transferred from individual to individual, cutting bureaucracy and costs in a substantial way. Crowdfunding and P2P lending have already existed before the pandemic (Agur et al. 2020; Lin et al. 2017), but the level of digital literacy acquired opens up for an unprecedented number of opportunities for targeted policy interventions aimed at putting together philanthropists with who was left behind during the last two years and small start-up projects with potential, but with little traditional creditworthiness with small venture capitalists without geographical constraints. Similar to dating apps, borrowers and lenders meet on matchmaking platforms (e.g. Estateguru, PeerBerry, Mintos and Lendermarket) that are often supported by Artificial Intelligence to optimize markets.

Artificial Intelligence is a powerful technology. And so far, we have only scratched the surface of what may be possible as more and more AI solutions are being developed. The potential of a learning algorithm able to understand and reinterpret complex human needs on several levels is to overcome behavioral attitudes that could impede the success of financial transactions without traditional intermediaries, mitigating the negative effects of economic shocks on social equality and building mutual trust in both the lender and the borrower. AI has been widely applied in the area of critical infrastructures, for example in public health, transport, water supply, electricity generation and telecommunications, increasing their resilience in case of extreme events. The same concept of resilience should be applied to financial assets of small and medium businesses and private citizens: in case of extreme events such as pandemics or other disasters, financial assets must be protected and able to recover quickly from such shocks. This could be achieved, among other solutions, by nudging small investors into choosing stocks with a higher ES (environmental and social) rating as they are shown to react less to volatility in the market (Albuquerque et al. 2020), by increasing digital and financial literacy among the population and by providing equitable access to technology, regardless of socio-economic status, ethnicity, gender, and age.

3.7   Conclusions

Wars, pandemics, terrorist attacks, and natural disasters are known to have a significant impact on financial markets. We provide an overview of the main economic and social shocks of the last century and how investors reacted in terms of financial decision-making. We discuss what such reactions entail in terms of social equality, and how these crises affect individual preferences such as the propensity toward risk-taking behavior, attitudes toward trust and confidence in the government.

A particular side effect of the recent Covid-19 pandemic is that of digitalization. We also dedicate part of our chapter to the phenomenon of digital literacy, which has experienced an unprecedented—yet forceful—surge after the first lockdown. While the costs of this process and the subsequent market instability are still unevenly distributed, digitalization represents a window to developing countries and classes with lower socio-economic status: it has the potential of being a remarkable opportunity for education, networks, economic growth and wealth redistribution. However, it is the task of global governments to actively create positive change and synergies, preventing them from ultimately worsening social and economic gaps among the population and between countries. This can be achieved by building up trust and investing in digital infrastructure and human capital, providing training and support to both private citizens and small businesses.

Social connectedness is a known predictor of upward social mobility and may act as a catalyst to the creation of a new global interconnectedness that can also be accessed by those who had previously been confined to an often limited local set of opportunities. Chetty et al. 2022a, 2022b study three types of connectedness based on social media friendships and ZIP codes. The first is between people with low and high socio-economic status, the second type is represented by social cohesion, and the third is defined by the level of civic engagement. They find the first type of (economic) connectedness to be the strongest determinant of upward income mobility. We claim that this holds to some extent also for global connectedness, regardless of physical proximity.

The present century is the century of complexity, as already predicted by a famous physicist and as we have seen, financial markets are no exception. The development of the concept of resilience is a consequence of this prophecy, as complexity cannot be solved with one pill. Recovering from the unforeseen—financially and socially—is however possible, with the right tools at hand. By increasing financial and digital literacy, investing in education and granting access to technology and by building up confidence in peers and in the government, a more equitable world can be created. Finally, the idea of *stakeholder capitalism* with companies acting with the common good in mind instead of immediate profits, together with the creation of a resilient, egalitarian, and environmentally sustainable system are the main pillars of *The Great Reset*, an initiative promoted by the World Economic Forum, which will be presented more indepth in the next chapter.

References

I. Agur, S.M. Peria, C. Rochon, Digital financial services and the pandemic: opportunities and risks for emerging and developing economies. Int. Monetary Fund **52**(5), 1–13 (2020)

D.A. Ahsan, Does natural disaster influence peoples risk preference and trust? an experiment from cyclone prone coast of Bangladesh. Int. J. Disaster Risk Reduct. **9**, 48–57 (2014). https://doi.org/10.1016/j.ijdrr.2014.02.005

G.A. Akerlof, R.J. Shiller, *Animal Spirits: How Human Psychology Drives the Economy, and Why it Matters for Global Capitalism* (Princeton University Press, 2010)

A.M. Al-Awadhi, K. Alsaifi, A. Al-Awadhi, S. Alhammadi, Death and contagious infectious diseases: impact of the COVID-19 virus on stock market returns. J. Behav. Exp. Financ. **27**, 100326 (2020). https://doi.org/10.1016/j.jbef.2020.100326

C.T. Albulescu, COVID-19 and the United States financial markets' volatility. Finan. Res. Lett. **38**(March 2020), 101699 (2021). https://doi.org/10.1016/j.frl.2020.101699

R. Albuquerque, Y. Koskinen, S. Yang, C. Zhang, Resiliency of environmental and social stocks: an analysis of the exogenous COVID-19 market crash. The Rev. Corporate Finan. Stud. **9**(3), 593–621 (2020). https://doi.org/10.1093/rcfs/cfaa011

J. Amankwah-Amoah, Z. Khan, G. Wood, G. Knight, COVID-19 and digitalization: the great acceleration. J. Bus. Res. **136**(April), 602–611 (2021). https://doi.org/10.1016/j.jbusres.2021.08.011

D. Arninda, D. Prasetyani, Impact of Covid-19 Pandemic: SMEs financing solutions through fintech lending education. In *Proceedings of the International Conference on Economics and Business Studies (ICOEBS 2022),* vol 655(Icoebs) (2022), pp. 25–31. https://doi.org/10.2991/aebmr.k.220602.004

A. Börsch-Supan, in *Survey of Health, Ageing and Retirement in Europe (SHARE) Wave 8. Release version: 8.0.0. SHARE-ERIC*. Norway—Data Archive and distributor of ESS data for ESS ERIC (2022). https://doi.org/10.6103/SHARE.w8.100

S.R. Baker, N. Bloom, S.J. Davis, K. Kost, M. Sammon, T. Viratyosin, The unprecedented stock market reaction to COVID-19. The Rev. Asset Pricing Stud. **10**(4), 742–758 (2020). https://doi.org/10.1093/rapstu/raaa008

W.B. Barrett, A.J. Heuson, R.W. Kolb, G.H. Schropp, The adjustment of stock prices to completely unanticipated events. Financ. Rev. **22**(4), 345–354 (1987)

J. Bernhofer, F. Costantini, M. Kovacic, Risk attitudes, investment behavior and linguistic variation. J. Human Resour. 0119–9999R2 (2021). https://doi.org/10.3368/jhr.59.2.0119-9999R2

P. Beutels, N. Jia, Q.Y. Zhou, R. Smith, W.C. Cao, S.J. De Vlas, The economic impact of SARS in Beijing, China. Tropical Med. Int. Health **14**(Suppl. 1), 85–91 (2009). https://doi.org/10.1111/j.1365-3156.2008.02210.x

D. Brounen, J. Derwall, The impact of terrorist attacks on international stock markets. Eur. Financ. Manag. **16**(4), 585–598 (2010)

T.R. Burch, D.R. Emery, M.E. Fuerst, What can "Nine-Eleven" tell us about closed-end fund discounts and investor sentiment? Financ. Rev. **38**(4), 515–529 (2003)

D.A. Carter, B.J. Simkins, The market's reaction to unexpected, catastrophic events: the case of airline stock returns and the September 11th attacks. Q. Rev. Econ. Finance **44**(4), 539–558 (2004)

A.H. Chen, T.F. Siems, The effects of terrorism on global capital markets. in *The Economic Analysis of Terrorism* (Routledge, 2007), pp. 99–122

M. Chesney, G. Reshetar, M. Karaman, The impact of terrorism on financial markets: an empirical study. J. Bank. Finance **35**(2), 253–267 (2011)

R. Chetty, M.O. Jackson, T. Kuchler, J. Stroebel, N. Hendren, R.B. Fluegge, S. Gong, F. Gonzalez, A. Grondin, M. Jacob, D. Johnston, M. Koenen, E. Laguna-Muggenburg, F. Mudekereza, T. Rutter, N. Thor, W. Townsend, R. Zhang, M. Bailey, N. Wernerfelt, Social capital I: measurement and associations with economic mobility. Nature **608**(7921), 108–121 (2022a). https://doi.org/10.1038/s41586-022-04996-4

R. Chetty, M.O. Jackson, T. Kuchler, J. Stroebel, N. Hendren, R.B. Fluegge, S. Gong, F. Gonzalez, A. Grondin, M. Jacob, D. Johnston, M. Koenen, E. Laguna-Muggenburg, F. Mudekereza, T. Rutter, N. Thor, W. Townsend, R. Zhang, M. Bailey, N. Wernerfelt, Social capital II: determinants of economic connectedness. Nature **608**(7921), 122–134 (2022b). https://doi.org/10.1038/s41586-022-04997-3

China's National Health Commission, in *Top expert: Disease spread won't be on scale of SARS* (2020). http://en.nhc.gov.cn/2020-01/21/c_75991.htm

T. Choudhry, September 11 and time-varying beta of United States companies. Appl. Finan. Econ. **15**(17), 1227–1242 (2005)

S. Corbet, C. Gurdgiev, A. Meegan, Long-term stock market volatility and the influence of terrorist attacks in Europe. Q. Rev. Econ. Finance **68**, 118–131 (2018)

S. Correia, S. Luck, E. Verner, Pandemics depress the economy. SSRN (2022)

Z. Da, J. Engelberg, P. Gao, The sum of all FEARS investor sentiment and asset prices. The Rev. Finan. Stud. **28**(1), 1–32 (2015)

K. Drakos, Terrorism-induced structural shifts in financial risk: airline stocks in the aftermath of the September 11th terror attacks. Eur. J. Polit. Econ. **20**(2), 435–446 (2004)

R. Eldor, R. Melnick, Financial markets and terrorism. in *The Economic Analysis of Terrorism* (Routledge, 2007), pp. 137–161

N. Engelhardt, M. Krause, D. Neukirchen, P. Posch, What drives stocks during the corona-crash? News attention versus rational expectation. Sustainability 12(12), 5014 (2020). https://doi.org/10.3390/su12125014

N. Engelhardt, M. Krause, D. Neukirchen, P.N. Posch, Trust and stock market volatility during the COVID-19 crisis. Finan. Res. Lett. **38**(August 2020), 101873 (2021). https://doi.org/10.1016/j.frl.2020.101873

B.S. Frey, S. Luechinger, A. Stutzer, Calculating tragedy: assessing the costs of terrorism. J. Econ. Surv. **21**(1), 1–24 (2007)

M. Glaser, M. Weber, September 11 and stock return expectations of individual investors. Rev. Finan. **9**(2), 243–279 (2005)

J.W. Goodell, COVID-19 and finance: agendas for future research. Finan. Res. Lett. **35** (2020). https://doi.org/10.1016/j.frl.2020.101512

L. Herbowski, The maze of the cerebrospinal fluid discovery. Anatomy Res. Int. **2013**, 1–8 (2013). https://doi.org/10.1155/2013/596027

M.T. Hon, J. Strauss, S.K. Yong, Contagion in financial markets after september 11: myth or reality? J. Finan. Res. **27**(1), 95–114 (2004). https://doi.org/10.1111/j.1475-6803.2004.00079.x

IMF, in *The Great Lockdown* (2020)

Italy's Council of Ministers, Comunicato stampa del Consiglio dei Ministri n. **31** (2020). http://www.governo.it/it/articolo/comunicato-stampa-del-consiglio-dei-ministri-n-31/14163

R.B. Johnston, O.M. Nedelescu, The impact of terrorism on financial markets. J. Finan. Crime (2006)

N. Kandasamy, B. Hardy, L. Page, M. Schaffner, J. Graggaber, A.S. Powlson, P.C. Fletcher, M. Gurnell, J. Coates, Cortisol shifts financial risk preferences. Proc. Natl. Acad. Sci. **111**(9), 3608–3613 (2014). https://doi.org/10.1073/pnas.1317908111

G.A. Karolyi, The consequences of terrorism for financial markets: what do we know? Available at SSRN 904398 (2006)

G.A. Karolyi, R. Martell, Terrorism and the stock market. Available at SSRN 823465 (2006)

M.R. Keogh-Brown, S. Wren-Lewis, W.J. Edmunds, P. Beutels, R.D. Smith, The possible macroeconomic impact on the UK of an influenza pandemic. Health Econ. **19**(11), 1345–1360 (2010). https://doi.org/10.1002/hec.1554

M.R. Keogh-Brown, R.D. Smith, The economic impact of SARS: how does the reality match the predictions? Health Policy **88**(1), 110–120 (2008). https://doi.org/10.1016/j.healthpol.2008.03.003

X. Lin, X. Li, Z. Zheng, Evaluating borrower's default risk in peer-to-peer lending: evidence from a lending platform in China. Appl. Econ. **49**(35), 3538–3545 (2017). https://doi.org/10.1080/00036846.2016.1262526

M. Mazur, M. Dang, M. Vega, COVID-19 and the march 2020 stock market crash. evidence from S&P1500. Finan. Res. Lett. **38**, 101690 (2021)

M. Miescu, R. Rossi, COVID-19-induced shocks and uncertainty. Europ. Econ. Rev. **139**, 103893 (2021). https://doi.org/10.1016/j.euroecorev.2021.103893

V. Niederhoffer, The analysis of world events and stock prices. The J. Business **44**(2), 193–219 (1971)

J. Nikkinen, S. Vähämaa, Terrorism and stock market sentiment. Financ. Rev. **45**(2), 263–275 (2010). https://doi.org/10.1111/j.1540-6288.2010.00246.x

S. Ramelli, A.F. Wagner, Feverish stock price reactions to COVID-19. Rev. Corp. Financ. Stud. **9**(3), 622–655 (2020). https://doi.org/10.1093/rcfs/cfaa012

M.Z. Sadique, W.J. Edmunds, R.D. Smith, W.J. Meerding, O. De Zwart, J. Brug, P. Beutels, Precautionary behavior in response to perceived threat of pandemic influenza. Emerg. Infect. Dis. **13**(9), 1307 (2007)

A. Sharif, C. Aloui, L. Yarovaya, COVID-19 pandemic, oil prices, stock market, geopolitical risk and policy uncertainty nexus in the US economy: fresh evidence from the wavelet-based approach. Int. Rev. Finan. Anal. **70**, 101496 (2020). https://doi.org/10.1016/j.irfa.2020.101496

F. Shear, B.N. Ashraf, M. Sadaqat, Are investor's attention and uncertainty aversion the risk factors for stock markets? International evidence from the COVID-19 crisis. Risks **9**(1), 2 (2020). https://doi.org/10.3390/risks9010002

R.M. Shelor, D.C. Anderson, M.L. Cross, Gaining from loss: property-liability insurer stock values in the aftermath of the 1989 California earthquake. J. Risk and Insurance 476–488 (1992)

C.U.M. Smith, E. Frixione, S. Finger, W. Clower, *The Animal Spirit Doctrine and the Origins of Neurophysiology* (Oxford University Press, 2012)

R.D. Smith, Responding to global infectious disease outbreaks: lessons from SARS on the role of risk perception, communication and management. Soc. Sci. Med. **63**(12), 3113–3123 (2006)

World Health Organization, Emergencies preparedness, response. Pneumonia of unknown origin—China (2020a)

World Health Organization, Novel coronavirus (2019-nCoV) Situation Report (2020b)

A. Worthington, A. Valadkhani, Measuring the impact of natural disasters on capital markets: an empirical application using intervention analysis. Appl. Econ. **36**(19), 2177–2186 (2004)

Juliana Bernhofer is an academic researcher at the "Marco Biagi" Department of Economics of the University of Modena and Reggio Emilia, honorary research fellow at the Ca' Foscari University of Venice and co-founder of CF Applied Economics, a center for applied economic analysis and research. She is an applied behavioral economist whose research interests span from the development of computational decision-support tools to empirical data analyses and human subject experiments in the areas of language economics, education and labor economics, taxation and health economics.

Anna Alexander Vincenzo is assistant professor at the Department of Economics and Management of University of Padua. Her current research interests involve financial accounting topics, with a particular focus on corporate tax avoidance, corporate governance, and the third sector.

Chapter 4
The AI's Role in the Great Reset

Federico Cecconi⊙

Abstract The phrase "Great Reset" is now in the public domain and appears in a number of analyses and discussions. However, we often remain ambiguous in this regard. It may be helpful to read the World Economic Forum's so-called white paper, Resetting the Future of Work Agenda in a Post-Covid World, to understand what theorists of this epochal maneuver are aiming for. This 31-page document describes how to run (or, as they say these days, implement) the software included in the CoViD-19 book. World Economic Forum founder Klaus Schwab and Thierry Malleret co-wrote The Great Reset (Schwab and Malleret in COVID-19: The Big Reset, 2020; Umbrello in The Journal Value Inquiry:1–8, 2021; Roth in European Management Journal 39.5:538–544, 2021). Why devote an entire chapter to this topic in a book about AI and its applications in finance? It's simple: updating the financial world to make it safer, more understandable, and more manageable necessitates extensive use of AI technologies. These technologies can be harmful up to a point.

Keywords The great reset · COVID-19 · Microtrends · Credit · ESG

4.1 To Summarize the Great Reset

The Great Reset could be the first great revolution with IT (AI as a main pillar) at its core: resetting the Future covers the decade from 2021 to 2030 (Fig. 4.1).

First and foremost, digitized work processes will be accelerated, with 84% of all work processes required to be digitized or made via video. To achieve complete social separation, approximately 83% of people will be required to work remotely, with no human interaction. At least half of all activities are expected to be automated, implying that direct human input, even in the same remote activity, will be drastically reduced, and upskilling and reskilling activities will also need to be computerized. Upskilling is the acquisition of new skills that allow a person to become more effective and qualified in their field. The acquisition of significantly different abilities in order

F. Cecconi (✉)
LABSS-ISTC-CNR, Via Palestro 32, 00185 Rome, Italy
e-mail: federico.cecconi@istc.cnr.it

F. Cecconi (ed.), *AI in the Financial Markets*, Computational Social Sciences,
https://doi.org/10.1007/978-3-031-26518-1_4

Fig. 4.1 Depicts the great reset, a World economic forum initiative

for a person to perform a different function is referred to as reskilling. In other words, the imperative in this scenario is to avoid human touch and achieve everything through computers, artificial intelligence, and algorithms.

Accelerate the implementation of retraining programs so that at least 35% of skills are "retrained," implying that the acquired skills must be abandoned.

Accelerate the reform of organizational structures. It is proposed that 34% of current organizational structures be "restructured," declaring them obsolete. The goal is to make room for new organizational frameworks in order to maintain complete control over all activities, including digitization.

Temporarily reassign people to different jobs: this is likely to affect around 30% of the workforce. This includes an examination of the pay scales.

Temporarily reduce the workforce: this fate is expected to affect 28% of the population. It is an unemployment issue because it is unclear what the phrase "temporary" means.

In addition, the Great Reset includes a credit plan in which a personal debt may be "forgiven" in exchange for the surrender of all personal assets to an administrative body or agency.

The previous COVID-19 spread is the most pressing reason to pursue a Great Reset. The pandemic is one of the deadliest public-health crises in recent memory, with hundreds of thousands of deaths already recorded. The war is far from over, with casualties still being reported in many parts of the world.

As a result, long-term economic growth, public debt, employment, and human well-being will all suffer. The Financial Times reports that global government debt has already reached an all-time high. Furthermore, unemployment is skyrocketing in many countries; for example, in the United States, one out of every four workers has filed for unemployment benefits since mid-March, with new weekly claims far exceeding historic highs. According to the International Monetary Fund, the global economy will contract by 3% this year, a 6.3 percentage point drop in just four months.

All of this will exacerbate existing climate and social crises. Some countries have already used the COVID-19 crisis to weaken environmental protections and enforcement. And discontent with social ills like rising inequality—the combined wealth of US billionaires has increased during the crisis—is growing.

In reality, there was already a requirement that can be summed up in one word: stability. We are currently witnessing a significant amount of skepticism in our societies and around the world. The main requirement, however, is stability. While the majority of this chapter will continue to concentrate on COVID and its consequences, it is important to remember that the need is much greater.

These crises, along with COVID-19, will worsen if not addressed, making the world less sustainable, unequal, and fragile. Incremental measures and ad hoc fixes will not suffice to avoid this scenario. Our economic and social systems must be completely rebuilt.

This necessitates unprecedented levels of collaboration and ambition. It is, however, not an impossible dream. One silver lining of the pandemic is that it has shown how quickly we can make radical lifestyle changes. Almost immediately, the crisis forced businesses and individuals to abandon long-held essential practices, such as frequent air travel and office work.

Similarly, populations have shown a strong willingness to make sacrifices for health-care and other essential workers, as well as vulnerable populations such as the elderly. And, in a shift toward the kind of stakeholder capitalism to which they had previously paid lip service, many corporations have stepped up to support their employees, customers, and local communities.

Clearly, there is a desire to create a better society. We must use it to secure the much-needed Great Reset. This will necessitate more powerful and effective governments, though this does not imply an ideological push for larger ones. And it will necessitate private-sector participation at every stage.

4.2 The Elements

The Great Reset agenda would consist of three major components. The first would steer the market toward more equitable outcomes. Governments should improve coordination (for example, in tax, regulatory, and fiscal policy), upgrade trade agreements, and create the conditions for a "stakeholder economy" to achieve this goal. Governments have a strong incentive to pursue such action in an era of shrinking tax bases and soaring public debt.

Furthermore, governments should carry out long-overdue reforms that promote more equitable outcomes. These may include changes to wealth taxes, the elimination of fossil-fuel subsidies, and new rules governing intellectual property, trade, and competition, depending on the country.

The second component of a Great Reset agenda would be to ensure that investments advance common goals like equality and sustainability. The large-scale spending programs that many governments are implementing represent a significant opportunity for progress in this area. The European Commission, for example, has proposed a €750 billion ($826 billion) recovery fund. The United States, China, and Japan all have aggressive economic stimulus plans.

Instead of using these funds, as well as investments from private entities and pension funds, to patch up the old system, we should use them to build a new one that is more resilient, equitable, and long-term sustainable. This includes, for example, the construction of "green" urban infrastructure and the creation of incentives for industries to improve their performance on environmental, social, and governance (ESG) metrics.

The third and final priority of a Great Reset agenda is to use Fourth Industrial Revolution innovations to benefit the public good, particularly by addressing health and social challenges. Companies, universities, and others banded together during the COVID-19 crisis to develop diagnostics, therapeutics, and possible vaccines; establish testing centers; develop mechanisms for tracing infections; and provide telemedicine. Consider what could be accomplished if similar coordinated efforts were made in every sector.

4.3 Microtrends

The Great Reset will entail a long and complex series of changes and adaptations at the micro level, that of industries and businesses. When confronted with this, some industry leaders and senior executives may be tempted to equate recovery with reboot, hoping to return to the old normal and restore what worked in the past: traditions, tried and true procedures, and familiar ways of doing things—in short, a return to business as usual. This will not occur because it cannot occur. In most cases, "business as usual" has died (or been infected with) COVID-19. Economic hibernation caused by blockades and social distancing measures has devastated some sectors. Others will struggle to recoup lost revenue before embarking on an increasingly narrow path to profitability as a result of the global economic downturn. However, for most businesses entering the post-coronavirus era, the key question will be striking the right balance between what worked in the past and what is required now to thrive in the new normal. For these businesses, the pandemic represents a once-in-a-lifetime opportunity to rethink their operations and implement positive, long-term change.

What will define the post-coronavirus business landscape's new normal?

How will businesses strike the best balance possible between past successes and the fundamentals now required to succeed in the post-pandemic era?

The response is obviously dependent and specific to each sector, as well as the severity of the pandemic. Aside from the few industries where companies will benefit on average from strong favorable winds (particularly technology, health and wellness), the journey will be difficult and at times treacherous in the post-COVID-19 era. For some industries, such as entertainment, travel, and hospitality, a return to pre-pandemic conditions is unthinkable in the near future (and perhaps never in some cases …). Others, particularly in manufacturing and food, are more concerned with adapting to the shock and leveraging new trends (such as digital) to thrive in the post-pandemic era. Size makes a difference as well. Small businesses face more challenges because they have smaller cash reserves and lower profit margins than larger businesses.

Most of them will have to deal with cost-to-income ratios that put them at a disadvantage in comparison to their larger competitors in the future. However, being small has some advantages in today's world, where flexibility and speed can make or break adaptation. Being agile is easier for a small structure than it is for a massive industrial structure.

That being said, regardless of their industry or the specific situation in which they find themselves, almost every single business decision maker around the world will face similar problems and will be required to answer some common questions and challenges. The following are the most obvious:

Should I encourage those who can (roughly 30% of the total US workforce) to work from home?

Will I reduce my business's reliance on air travel, and how many face-to-face meetings can I significantly replace with virtual interactions?

How can I make the business and our decision-making process more agile, allowing us to move faster and more decisively?

How can I accelerate digitization and digital solution adoption?

We are still in the early stages of the post-pandemic era, but there are already new or accelerating trends at work. For some industries, this will be a boon; for others, it will be a major challenge. However, it will be up to each company to capitalize on these new trends by adapting quickly and decisively across all industries. Companies that demonstrate the most agility and flexibility will emerge stronger.

4.4 Digitization is Being Accelerated

The buzzword of "digital transformation" was the mantra of most executive boards and committees prior to the pandemic. Digital was "critical," it had to be implemented "decisively," and it was regarded as a "prerequisite for success"! Since then, the mantra has become a must, even in the case of some businesses, a matter of life and death. This is simple to explain and comprehend. During our incarceration, we were completely reliant on the Internet for everything, from work and education to socializing. It is online services that have allowed us to maintain a semblance of normalcy, and it is natural that "online" has benefited the most from the pandemic, providing a huge boost to technologies and processes that allow us to do things remotely, such as universal Internet to broadband, mobile and remote payments, and viable e-government services, among others. As a result, businesses that already operate online will have a long-term competitive advantage. Businesses in industries as diverse as e-commerce, contactless operations, digital content, robots, and drone deliveries (to name a few) will thrive as more and more things and services are delivered to us via our cell phones and computers. It is no coincidence that companies such as Alibaba, Amazon, Netflix, and Zoom have emerged as "winners" as a result of the lockdown.

In general, the consumer sector was the first and fastest to move. From the required contactless experience imposed on many food and retail businesses during lockdowns to virtual showrooms in the manufacturing industry that allow customers to browse and select the products they prefer, most business-to-consumer companies have quickly recognized the need to provide their customers with a digital journey "from start to finish."

With the end of some lockdowns and the recovery of some economies, similar opportunities arose in business-to-business applications, particularly in the manufacturing sector, where physical distancing rules had to be implemented quickly, often in difficult environments. (For example, an assembly line.)

As a result, the Internet of Things has made significant progress. Some companies that were slow to adopt IoT prior to the lockdown are now embracing it in large numbers with the specific goal of doing as many things remotely as possible. All of these different activities, such as equipment maintenance, inventory management, supplier relationships, and security strategies, can now be performed (to a large extent) by a computer. The Internet of Things provides businesses with the ability to not only execute and comply with social distancing rules, but also to reduce costs and implement more agile operations.

Because of the inherent fragility of global supply chains, discussions about shortening have been ongoing for years. They are often intricate and difficult to manage. They are also difficult to monitor in terms of environmental and labor law compliance, potentially exposing companies to reputational risks and brand damage. In light of this troubled history, the pandemic has put the final nail in the coffin of the principle that companies should optimize supply chains based on the cost of individual components and on a single source of supply for critical materials. In the post-pandemic era, "end-to-end value optimization"—a concept that includes resilience, efficiency, and cost—will be dominant. The formula states that "just-in-case" will eventually replace "just-in-time."

The shocks to global supply chains discussed in the macro section will affect both large and small businesses. But, in practice, what does "just in case" mean? The globalization model conceived and built at the end of the last century by global manufacturing companies in search of low-cost labor, products, and components has reached its limits. It has fragmented international production into ever more intricate fragments, resulting in a system managed on an intimate scale that has proven extremely lean and efficient, but also extremely complex and, as a result, extremely vulnerable (complexity brings fragility and often provokes instability). As a result, simplification is the antidote, which should result in increased resilience. As a result, the "global value chains," which account for roughly three-quarters of all global trade, will inevitably decline. This decline will be exacerbated by the new reality that companies that rely on complex just-in-time supply chains can no longer assume that the World Trade Organization's tariff commitments will protect them from a sudden wave of protectionism from somewhere part.

As a result, they will be forced to prepare for a prolonged outage by reducing or localizing their supply chain and devising alternative production or supply plans. Any business whose profitability is dependent on the just-in-time global supply chain principle will have to rethink how it operates and will almost certainly have to sacrifice the idea of maximizing efficiency and profits for the sake of "supply security" and resilience. Resilience will then become the primary consideration for any business serious about protecting itself from disruption, whether that disruption is caused by a specific supplier, a potential change in trade policy, or a specific country or region. In practice, this will force businesses to diversify their supplier base, even at the expense of stockpiling and creating redundancy. It will also compel these companies to ensure that the same is true within their own supply chain: they will assess resilience throughout the entire supply chain, all the way down to their final supplier and possibly even their supplier's suppliers. Costs of production will inevitably rise, but this will be the cost of building resilience. At first glance, the automotive, electronics, and industrial machinery sectors will be the most affected because they will be the first to change production models.

4.5 ESG and Stakeholder Capitalism

The fundamental changes that have occurred in each of the five macro categories examined in Chap. 1 over the last decade or so have profoundly changed the environment in which companies operate. They have increased the importance of stakeholder capitalism and environmental, social, and governance (ESG) considerations in creating long-term value (ESG can be considered the benchmark for stakeholder capitalism) (Puaschunder 2021).

The pandemic struck at a time when many issues, from activism to climate change and growing inequalities to gender diversity and #MeToo scandals, had already begun to raise awareness and increase the importance of stakeholder capitalism and ESG considerations in today's interdependent world. Whether they are openly married or not, no one can deny that the fundamental purpose of businesses can no longer be the unbridled pursuit of financial profit; they must now serve all of their stakeholders, not just those who hold shares. This is supported by early anecdotal evidence pointing to an even brighter future for ESG in the post-pandemic era. This is explained in three ways:

1. The crisis will have instilled or reinforced a strong sense of responsibility and urgency in most ESG issues, the most important of which is climate change. Others, however, such as consumer behavior, the future of work and mobility, and supply chain responsibility, will rise to the top of the investment agenda and become an essential component of due diligence.

2. The pandemic has shown that the absence of ESG considerations has the potential to destroy significant value and even threaten a company's profitability. After that, ESG will be more fully integrated and internalized into a company's core strategy and governance. It will also alter investor's perceptions of corporate governance. Tax returns, dividend payments, and wages will be scrutinized more closely for fear of incurring a reputational cost if a problem arises or becomes public.
3. Promoting employee and community goodwill will be critical to improving a brand's reputation.

4.6 And Now for the Big Reveal...

As a result, if the world chose to go down this road, defend and pursue these goals, and even partially embrace these theses, we would almost certainly end up in a situation where AI algorithms would manage what we simply cannot manage without them. A truly historic revolution. When discussing digitization, many people focus on the more immediate consequences, such as job loss. On the contrary, the opposite thesis is supported here for the first time: intelligent IT technologies will be able to assist us in managing complexity. A truly intriguing challenge: AI in exchange for stability (Da Silva Vieira 2020; Murmann 2003).

References

K. Schwab, T. Malleret, COVID-19: The Big Reset (2020)

S. Umbrello, "Should we start over? 'COVID-19: the great reset' ed. by K. Schwab, T. Malleret is reviewed". J. Value Inquiry 1–8 (2021)

S. Roth, The great paradigm shift in management and organizational theory. A European point of view. Europ. Managem. J. **39.5**, 538–544 (2021)

J.M. Puaschunder, Environmental, social, and corporate governance (ESG) diplomacy: a corporate and financial social justice great reset is required. in *The Proceedings of the 23rd Research Association for Interdisciplinary Studies (RAIS) Conference* (2021) pp. 119–124

M.M. Da Silva Vieira, School culture and innovation: does the COVID-19 post-pandemic world invite transition or rupture?. Europ. J. Soc. Sci. Educ. Res. **7.2**, 23–34 (2020)

J.P. Murmann et al., Evolutionary thought in management and organization theory at the turn of the millennium: a symposium on the state of the art and future research opportunities. J. Managem. Inquiry **12.1**, 22–40 (2003)

Federico Cecconi is R&D manager for QBT Sagl (https://www.qbt.ch), responsible for the LABSS (CNR) of the computer network management and computational resources for simulation, and consultant for Arcipelago Software Srl. He develops computational and mathematical models for the Labss's issues (social dynamics, reputation, normative dynamics). For Labss performs both the dissemination that training. His current research interest is twofold: on the one hand, the study of socio-economic phenomena, typically using computational models and databases. The second is the development of AI model for fintech and proptech.

Chapter 5
AI Fintech: Find Out the Truth

Federico Cecconi◉ and Alessandro Barazzetti◉

Abstract We propose one example of application that could have a large impact on our economies: the use of artificial agents to reproduce dynamics, case study the spreading of fake news. It is known that a large number of applications (some described here) are dedicated to investing in financial markets by observing the prevailing trends, often carrying out operations at very high speed. These applications often cause problems by accelerating downtrends irrationally. The markets are running for cover against this type of problem. But automatic traders are emerging who are able to evaluate not the trends but directly the news that alone is at the origin of the trends themselves. This poses the following problem: if cultural diffusion (in this case news, but it could be posts on social networks, little changes) is made problematic by the appearance of 'fake.news', how can we be calm about the functioning of our automatic traders? Here we propose a completely new approach to this problem, which makes use of a widespread AI technology, artificial agents. These agents reproduce the phenomenon of spreading fake news and give indications on how to fight it (Chen and Freire in Discovering and measuring malicious URL redirection campaigns from fake news domains, pp. 1–6, 2021; Garg et al. in Replaying archived twitter: when your bird is broken, will it bring you down? pp. 160–169, 2021; Mahesh et al. in Identification of Fake News Using Deep Learning Architecture, pp. 1246–1253, 2021).

Keywords Fake news · Automatic trading · Agent based simulation

F. Cecconi (✉)
LABSS-ISTC-CNR, Via Palestro 32, 00185 Rome, Italy
e-mail: federico.cecconi@istc.cnr.it

A. Barazzetti
QBT Sagl,, Via E.Bossi, 4 CH6830 Chiasso, Switzerland
e-mail: alessandro.barazzetti@qbt.ch

© The Author(s), under exclusive license to Springer Nature Switzerland AG 2023   65
F. Cecconi (ed.), *AI in the Financial Markets*, Computational Social Sciences,
https://doi.org/10.1007/978-3-031-26518-1_5

5.1   The Diffusion of Culture

The sociological literature relating to the dynamics of diffusion and consolidation of culture is very extensive. Here we have chosen to start from the concept of culture defined by Axelrod who sees the dissemination of culture as the output of a process of reciprocal influence between social actors who share specific values and distinctive traits. In fact, similar agents will tend to prefer interaction between them and consequently a circle of homophilia growth is created between them. In this context, the process of interaction between agents plays a fundamental role.

In particular, the omnipresence of social structures can be found within the daily behavior of social actors to the point of co-occurring in the determination of their own identity. In the collective dimension, the role of social media is fundamental in determining collaborative digital networks that can be "weak" and instantaneous like affinity networks but can also structure themselves to the point of constituting themselves as real social groups.

Therefore, within the dimension of social media, the relational potential acquires such importance that the consumer is not only a user of messages and contents but also becomes their producer and editor, generating a continuous circle of knowledge.

In this interpretative frame the figure of the prosumer is born, i.e. the one who is the producer (producer) and user (consumer) of the media product. The term prosumer was coined in 1980 by the American sociologist Toffler already in 1975 by Jean Cloutier had elaborated a similar idealtype defining it self-media, in both cases we refer to the ability of the subject to be producer and user in a contemporary way of cultural and media products.

The digital prosumer, however, is not to be considered as a particular one-off figure but rather defines any user of the network. Since each digital actor is also a content producer, in particular a data producer. The action of producing a digital content within the web space, in fact, is not necessarily limited to the creation of a specific online multimedia content (post, video, image) it also concerns simpler and apparently low communicative activities such as for example share or retweet content, insert a like or geolocate yourself in a specific place.

Choosing to open one website rather than another also means that you have also unknowingly contributed to its management.

5.2   Research Design

The aim of this AI solution is to reconstruct the process by which conspiracy-themed articles are shared by Facebook pages forming a bubble of news from unofficial sources. In particular, it was coded a simulative model based on artificial agents. The agents are articles, social pages, websites and users; these interacting with each other make up the pages-to-articles network ('pages to articles' network, pta network), which is the focus of our investigation (Gupta and Potika 2021; An et al. 2021; Yu et al. 2021; Singh and Sampath 2020).

The simulation project is based on an exploratory survey structured through two research issues.

Issue 1—Some misleading information contents, present in the digital space, have some properties that make them desirable for specific Facebook pages and groups and based on specific characteristics these contents spread within the Social Network. The properties concern respectively: the conspiracy-based content, the intensity of the conspiracy content, the political orientation, incitement to hatred and incitement to fear.

Issue 2—Concerns the analysis of the impact of the content of the articles on the topology of the eco chamber through the simulation model.

5.3 Data Mining

In the first phase, an analysis of digital footprints was carried out with respect to Italian counter-information sources identified through a qualitative observation phase of the interactions and content shared within a set of public groups on Facebook. In this regard, 197 articles have been selected on the basis of their content in line with the typical themes of the conspiracy movement (Choudhary and Arora 2021; Garg and Jeevaraj 2021).

Subsequently, the interactions of each article within Facebook were detected through the Social Network Analysis tool. This phase was important both to circumscribe the ecosystem of conspiracy-minded public pages and groups and to identify the posts with the most interactions. Some empirical evidence emerges from the preliminary analysis: (1) there are Facebook sources that only share conspiracy-based content; (2) these articles to be of interest with respect to the pages present specific contents; (3) the contents found come in turn from sources specialized in the dissemination of conspiracy information.

The set of data collected and the processing carried out made it possible to identify the main units (agents) at the basis of the construction process of the echochamber:

Websites or the sites (blogs, information portals, online newspapers, etc.) that produce and contain the articles identified.

Articles, the articles collected in the first phase of the research which have the characteristic of presenting contents of information.

Pages, this category includes the pages and public Facebook groups that have shared the articles.

Users, users who interacted within Facebook with the articles.

5.4 ABM Simulation

The purpose of the simulation is to reconstruct the pages versus articles. In the simulation we have to create links between pages and articles. This must be done using the properties of the articles, websites, pages and users taken from the survey. In the start phase, the links between pages and articles are not loaded since that type of network represents the output of the simulation.

On the basis of the data collected in the Data Mining phase, the simulation model was designed and codified in two operational tasks: definition of the agents, and definition of the algorithm. The data collected was used to build the artificial agents, divided into four agent sets: Articles, Websites, Pages and Users (Fig. 5.1).

In detail, the four types are structured as follows: the web sites are connected only with the articles, the articles are connected with the Web sites and with the Pages, the pages are connected with the articles and with the users and finally the users are connected exclusively with pages.

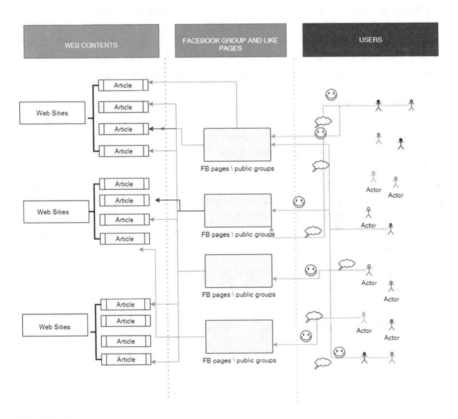

Fig. 5.1 The agentsets

Specifically, the Articles dataset was constructed by inserting all the articles identified in the first phase of the research classified in their content in relation to the following interpretative categories: (1) contents, the presence / absence of conspiratorial content; (2) intensity, intensity of the content on a scale from 1 to 5; (3) Issue, or the conspiracy narratives present in the article; (4) political, if the article is politically aligned; (5) political orientation, what political orientation does it present; (6) incitement to hate, if it presents characteristics of incitement to hate; (7) incitement to fear, incitement to fear.

The Websites dataset was composed by inserting the following information for each website identified: sum of interaction, count of articles, fame index and exclusivity. Sum of interaction indicates the total sum of the interactions that the contents from that site have had within Facebook, while Count of articles is a value that indicates the number of articles collected from that website. These two values together make up Fame index, that is a synthetic index that expresses when a website is popular within the detected echo chamber.

The Pages database consists of the following items: Source or the name of the Facebook page or public group, Follower, the number of followers of the page, Interaction or the interactions collected for each article.

Finally, User is made up of User Id or a label that identifies the user, Content, the content of the product comment for each page.

5.5   Result: The AI Anti 'Fake News' Machine

In Fig. 5.2 we show the interface of the simulator.

The simulator, as already mentioned, loads a repertoire of articles, web pages on which they could be published, users who produce these contents and web sites where the articles can be found. Once the upload has been completed, the simulator hypothesizes how the articles will connect to the pages, what is important or what is less so that homogeneous groups are created (the example used here concerns homogeneous groups of health conspiracy theorists).

In the image we used, the yellow figures are the agents, the houses are the websites, and the top two boxes contain the articles and web pages where they have been cited. Obviously the entire interface is selectable, cross-queries can be made, for example to find out which factors are decisive in creating groups where all users are in contact (Fig. 5.3). The simulator receives a certain number of parameters as input, such as how much weight it is necessary to give to the presence or absence of incitement to violence within the articles.

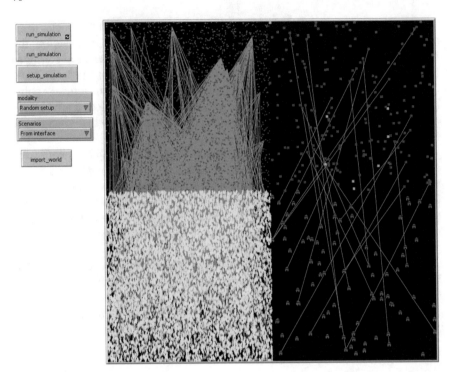

Fig. 5.2 The interface of NFN System

The main result of the entire simulation is a report summarizing the factors that have led (or not) to the formation of homogeneous groups of conspiracy theorists (an example in Table 5.1). At the end of the simulation, the report can be used to select the contents to feed to the automatic asset evaluation systems, so as to be able to say for example … *avoid contents where there is an <u>apparently correct evaluation of the effect of vaccinations</u> with RNA; because these evaluations are at the basis of the creation of conspiracy bubbles…*, which can mislead the automatic evaluation systems.

In other words, NFN System[1] (or other simulators based on artificial agents of the same type) can function as risk 'selectors', indicating which semantic aspects, extracted in this case from social networks, but similarly extracted from news, can constitute a risk from the point of view of those who use those contents in automatic systems based precisely on semantics (Marium and Mamatha 2021; Nath et al. 2021; Uppal 2020).

---

[1] It is possible to have a demo, customizations, or even a complete suite of news evaluation products from QBT Sagl, https://www.qbt.ch.

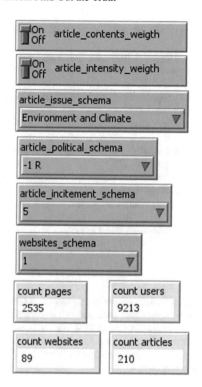

Fig. 5.3 Some input parameters of the simulator

Table 5.1 An extract of an NFN System evalutaion report

	Factors
Fake news bubble	RNA vaccine distribution
	RNA production
	Vaccine (secondary)
Groups	Use of 'hate' in main issue
	Use of 'science citation' in main issue
	Use of 'UE' in secondary issue

References

Z. Chen, J. Freire, Discovering and measuring malicious URL redirection campaigns from fake news domains (2021), pp. 1–6

A. Choudhary, A. Arora, ImageFake: an ensemble convolution models driven approach for image based fake news detection (2021) pp. 182–187

K. Garg, H.R. Jayanetti, S. Alam, M.C. Weigle, M.L. Nelson, Replaying archived twitter: when your bird is broken, will it bring you down? (2021), pp. 160–169

R. Garg, S. Jeevaraj, Effective fake news classifier and its applications to COVID-19 (2021), pp. 1–6

K. Gupta, K. Potika, Fake news analysis and graph classification on a COVID-19~twitter dataset (2021), pp. 60–68

K.A., Han, D. Damodharan, L. Anupam, K. Manoj Sagar, K. Murugan, S.S. Ajay, Discover pretend disease news misleading data in social media networks using machine learning techniques (2021), pp. 784–788

R. Mahesh, B. Poornika, N. Sharaschandrika, S.D. Goud, P.U. Kumar, Identification of Fake News Using Deep Learning Architecture (2021), pp. 1246–1253

A. Marium, G.S. Mamatha, BERT model for classification of fake news using the cloud processing capacity (2021), pp. 1–6

K. Nath, P. Soni, Anjum, A. Ahuja, R. Katarya, Study of fake news detection using machine learning and deep learning classification methods (2021), pp. 434–438

S. Yu, F. Xia, Y. Sun, T. Tang, X. Yan, I. Lee, "Detecting outlier patterns with query-based artificially generated searching conditions". IEEE Trans. Comput. Soc. Syst. **8**(1), 134–147 (2021)

M. Singh, N. Sampath, Truth discovery algorithm on big data social media tweets (2020), pp. 1553–1561

A. Uppal, V. Sachdeva, S. Sharma, Fake news detection using discourse segment structure analysis (2020), pp. 751–756

Federico Cecconi is R&D manager for QBT Sagl (https://www.qbt.ch), responsible for the LABSS (CNR) of the computer network management and computational resources for simulation, and consultant for Arcipelago Software Srl. He develops computational and mathematical models for the Labss's issues (social dynamics, reputation, normative dynamics). For Labss performs both the dissemination that training. His current research interest is twofold: on the one hand, the study of socio-economic phenomena, typically using computational models and databases. The second is the development of AI model for fintech and proptech.

Alessandro Barazzetti aeronautical engineer at the Politecnico di Milano, Professor at Ludes Off-Campus Semmelweis University and pilot is the owner of QBT Sagl, a company operating in the development of Algorithms and Software in the Fintech and Proptech sector and of Science Adventure Sagl, a company specialized in the development of applied AI solutions to Health and Care and advanced projects in the Aerospace sector. Alessandro is also founder of SST srl, a company devoted to sustainability studies and certification and Reacorp Sagl a company devoted to Fintech and Proptech services.

Chapter 6
ABM Applications to Financial Markets

Riccardo Vasellini

Abstract Traditional economic theories struggle to explain financial market volatility and the macroeconomy. Traders seldom assume the role of rational optimizers, as described in standard economic models, in financial markets. Boom and bust cycles are widespread, as are firm values based purely on the flashy personalities of CEOs, reckless speculators, and a general sense of irrationality. Clearly, this is not enough to condemn old economic theories as "useless." In reality, knowing that human behavior leads in self-regulating, stable markets where prices never deviate too far from equilibrium can provide exciting and practical outcomes Instead of evaluating the financial system as a whole and modeling it from the top down, one might examine its agents and see if system-wide characteristics emerge from their interaction. This is extensively used to investigate so-called Complex Systems, which exhibit significant non-linearity and unstable equilibrium states that are easily disrupted by little shocks. Traditional economic theories are effective in illustrating a relatively narrow system state, namely one in which the market is in equilibrium. Alternative techniques, however, backed by complex system theory, may allow for the portrayal of a more generic model in which equilibrium states are a subfield of a larger model. ABM is a valuable approach for simulating complex systems. Bottom-up modeling does not need scientists to approximate the entire system to a differential equation. They could instead mimic the interactions of the system's single agents. If the agents' behavioral assumption is right, shared behaviors should "emerge" in the system.

Keywords Agent based modeling · Financial markets · Cognition

6.1 Introduction

Traditional economic theories have a particularly tough time explaining financial market fluctuations and how they relate to the macroeconomy (LeBaron 2000). In the financial markets, traders seldom ever play the part of the rational optimizers

R. Vasellini (✉)
Università di Siena, Siena, Italy
e-mail: r.vasellini@student.unisi.it

© The Author(s), under exclusive license to Springer Nature Switzerland AG 2023
F. Cecconi (ed.), *AI in the Financial Markets*, Computational Social Sciences,
https://doi.org/10.1007/978-3-031-26518-1_6

that are frequently depicted in traditional economic models, in fact is not uncommon to observe boom and bust cycles, company valuations depending only on the glitzy personalities of CEOs, reckless speculators, and a great deal of irrationality overall.

Obviously, this not suffice to disregard traditional economic models as "useless", in fact thinking that human behavior results in self-regulating, stable marketplaces where prices never stray too far from equilibrium, can yield interesting and realistic results (Iori and Porter 2012).

However, an alternative approach is possible: instead of evaluating the financial system as a whole, and model it using a top-down approach, one might think to study it by understanding the agents populating the system and observe if properties common to the system emerge from their interaction. This is often the approach used to study the so-called Complex Systems, i.e., systems exhibiting high degrees of non-linearity, in which equilibrium states are unstable and small perturbations easily disrupts them.

One could argue that while traditional economic theories are successful at illustrating a very specific system state, namely one in which the market is in equilibrium, alternative approaches, supported by the theory of complex systems, may enable the illustration of a more general model in which the specific instances of equilibrium states are a subfield of broader model.

A very good way to model complex system is to use Agent Based Modelling. In this type of modelling, the bottom-up approach is king and scientists don't have to concern themselves with approximating the whole system to a differential equation. Instead, they may concentrate their efforts on simulating the interactions of the single agents that are intended to exist in the system. If the behavioural hypothesis upon which the agents have been created is sound, one expects to see the "emergence" of behaviours that are shared by the entire system.

To those not too familiar with this type of modelling imagine the following:

You want to describe how a balloon filled with gas behave, instead of writing down the Gay-Lussac's law, $PV = nRT$, thus using a top-down approach, you describe how a single atom or molecule of the gas behaves when it exchanges heat, modify is quantity of movement or gets hit by another atom of the gas. Then you use a computer to simulate all these interaction among atoms and other atoms and atoms and the environment. If you did a good job the results emerging will be coherent with the Gay-Lussac's law even though you could have known nothing about the balloon as a whole.

Obviously, in the specific case of this metaphor, it seems pretty counterintuitive and unnecessarily complicated to use such approach. However, there are many circumstances when it might be difficult, if not impossible, to use conventional mathematical models. When dealing with a large number of equations and several heterogeneous variables with a lot of indices, mathematical models are often not particularly effective.

These compliances can, instead, be managed better if described by algorithms and iterative processes. In particular when parts of the system interact frequently among each other's leading to the emergence of important features, to understand the system is then usually very hard to implement traditional mathematical models. What I just wrote is not entirely accurate, what would be accurate is to say that when the system as whole cannot be analyzed clearly by the scientists trying to describe it, then it's very hard to implement traditional models. When a bird's eye view is not possible, like in the case of societies or economies where even the people studying the system are part of it, then it's useful to try using an ABM process, starting from the bottom, from what we can understand and describe in a stylized way for the purpose of a simulation. This is the true power of ABM. To use another metaphor, it's much easier to describe an ant, a very simple living organism, then to describe the entire ants' nest, whose behavior cannot be comprehended by simply describing that of a single ant. However, harnessing the power of simulation is possible, by setting simple behavioral rules for the single ant, describe the whole nest.

Now that we've established what ABM is all about, we can provide some modelling overview and, following that, we will present a novel model developed by the writers of this book.

6.2 ABM

ABMs present a challenge to the neoclassical idea that agents generate rational expectations based on the assumption that they have full understanding of the economy and limitless computing capacity (Iori and Porter 2012). The all-knowing rational agent is substituted by one with bounded rationality (Simon 1997), able to infer about the markets thanks to a combination of behavioural rules and heuristics. In fact, we can use agents' modelling types to classify different modelling approaches:

Zero Intelligence Agents: Agents are scarcely, or not at all, able to learn and that their behavioural rules are extremely simple. The point of these type of models is to observe if features actually present in real financial markets can emerge from the system's structure. A very good example of this is the model developed by Farmer et al. (2005). The authors take the notion of a non-rational agent to its final extreme, creating agents that are completely devoid of rationality. In the model the agents simply place orders randomly (precisely orders arrive as a Poisson random process) and participate to a continuous double auction, i.e., they can place buy or sell orders at any time they want. Therefore, the model drops any ambition of describing market strategies or human behaviour and instead focuses mostly on characterising the limitations that the continuous double auction imposes on the system. The sole difference between agents is how frequently they place orders: impatient agents will be forced to place a fixed number of orders per unit time, whereas patient agents will put orders at random, not just for pricing but also for time. The authors then compared and evaluated their findings against real data from the London Stock Exchange Market, explaining approximately 96% of the spread and 75% of the variance in the price diffusion rate using a single free parameter.

What should be surprising is that just by making very simple assumption, in this case the simplest possible, on the behaviour of the agents it's possible to reproduce market's and prices' features by focussing the modelling attention on how the agents actually interact (continuous double auction).

In traditional models a common feature that traders lack is the possibility of being chartists. Given the rational assumption the only possibility for the stock trader is to be a fundamentalist always trying to discover the real valuation of a stock and trade consequently.

However, when implementing agents more complex than the zero-intelligence ones it's possible to include as many behavioural rules as wanted. Within this framework traders can pursue different strategies and react differently to signals, either from the market or from other agents. Iori and Porter (2012) differentiate between heterogenous agents whose interaction is mediated by the market and those who can interact directly among each other's. The latter type can has been useful to explain herding behaviour in finance (Hirshleifer and Slew 2003), while the former has been useful to understand the role of learning (LeBaron 2000).

6.3   Our Model

We started developing this model starting from a fundamental idea: Markets are diverse ecosystems inhabited figuratively by a myriad of different species, each with their own set of interests and ambitions. Understanding how these groups of people act and interact might help us understand market behaviour.

This is precisely what Andrew Lo argues in his Adaptive Market Theory, which is an attempt to incorporate the Efficient Market Theory into a broader framework in which it becomes a special case. Considering financial markets in this manner entails conceiving of them as complex systems; thus, to understand them, we must employ the methods we use to examine complexity. We decided that the best approach would have been to use AB Modelling.

Moreover, we started wondering if these "groups" living in the financial ecosystem could emerge spontaneously from the interaction of multiple agents trading in the market. By "groups," we refer to clusters of agents exhibiting dynamically similar behaviour. In a genuine market, this may include institutional and retail investors or algorithmic and manual traders, to name a few examples. The groups are limited just by the person's defining them imagination. We decided to focus only on two types of investors: Fundamentalists, and Chartists (whom we, unfortunately, called "Trendists").

Our primary objective was to demonstrate if and how agents without any a-priori configuration or bias toward being fundamentalists or chartists, nor any explicit external trigger coming from the system pushing in one direction or the other, could cluster into groups of investors whose behaviour follows a particular pattern over time. These behaviours emerge spontaneously only through the two mechanisms of adaptation available to the agents, i.e., learning and imitation. Now we will proceed to describe the model into more details.

6.3.1 Overview

Each day, the traders will analyse each of the three stocks using two distinct methodologies. They do a fundamental examination of the stock, simulating the act of investigating the worth of the underlying corporation that issued the shares, and consequently price the stock. They will also do a trend analysis, focusing solely on the previous performance of the stock while neglecting the underlying asset. The sequence in which these two pricing methods are executed is unimportant. Then, they will weigh the two acquired prices based on their current market outlook, or whether they feel it is preferable to be "fundamentalists" or "chartists." In addition, the price will be affected by how greedy or fearful market participants are at the moment. Finally, following these analyses, they will determine which stocks to sell or purchase and in what quantities.

At this point orders are inserted in the ledger.

The market will then match orders. It will look for a purchase (if the order is a sell) or a sell (if the order is a buy) order that matches the desired price and quantity, beginning with the oldest order issued. Price and quantity do not need to be identical; an agent might purchase fewer stocks at a cheaper price than he initially desired and then purchase the remainder from another agent. The day concludes when all orders that might be fulfilled have been finished. Then the agents will learn, which entails upgrading their decision-making heuristics depending on whether they are better off than before and how they may have been had they made other judgments. They will then emulate agents with whom they are associated who are wealthier than they are, adjusting their heuristics appropriately. Every two weeks, the fundamental value of the stock will vary by a modest proportion. This is done to imitate actual changes in the assets of the underlying firms. Average stock prices and other agents' fundamental properties will be preserved. At this moment, the cycle may resume.

6.3.2   Agents and Stocks Features

Agents behave following three main decisional parameters; Greediness, Truesight and Trendsight. Moreover, they have a certain number of resources and a total net worth. More details on these 5 features are in Table 6.1.

At the beginning of the simulation every agent is initialized attributing each of its a random value from a certain distribution depending on the feature.

Table 6.1 Agent properties

Resources	Liquidity of the agents, money at their disposal. It is initially distributed in the population as a Pareto distribution with $\alpha = 1.16$
Net worth	Total wealth, sum of resources and stock owned times their value at the moment
Greediness	Willingness to be in the market. Value between 0 and 1, 0 being extreme fear of the market and 1 extreme F.O.M.O. from the market. It is distributed in the population as a Normal distribution $N \sim (50, 20)$
Truesight	Ability to perform fundamental analysis of a stock. Value between 0 and 1, 0 being total inability to price a stock based on its fundamentals and 1 ability of perfectly doing it. It is initially distributed in the population as a Normal distribution $N \sim (50, 20)$
Trendsigth	Importance given by an agent to trend analysis of a stock. Value between 0 and 1, 0 being complete indifference to trends while 1 is the opposite. It is initially distributed in the population as a Normal distribution $N \sim (50, 20)$

Table 6.2 Stock properties

True value	The price value one would obtain from a perfect price analysis, it can be seen as the real value of a stock, not perfectly known by anyone. It changes each 25 ticks, growing or decreasing by values included between 0.5 and 6%
Trend value	It's simply given by a weighted average of the last 5 prices $$\text{Trend Value} = 0.4\,\text{Avg.Price}_n$$ $$+\,0.25\,\text{Avg.Price}_{n-1}$$ $$+\,0.15\,\text{Avg.Price}_{n-2}$$ $$+\,0.1\,\text{Avg.Price}_{n-3}$$ $$+\,0.1\,\text{Avg.Price}_{n-4}$$ Where n is the tick value

These 5 features, and in particular the 3 decisional one, are used by each agent to take a decision on which stock buy or sell and at what price.

The stocks present on the market are 3, they differ for their initial price and how much one of their 2 fundamental proprieties (the "true value") variates in time.

The 2 fundamental proprieties are better explained in Table 6.2.

6.3.3 Mapping Process

At each simulation iteration, the features of each agent's order are determined by the three decisional parameters of the agents and the two characteristics of the stocks. In practise, an agent will establish a price as if it were only behaving as a Fundamentalist, thus using only its "truesight," a price as if it were only behaving as a Chartist, thus using only its "trendsight," and then a total price weighting both prices based on their trendsight and truesight values. The agent will then determine whether to buy or sell based on his own "greediness," selling when greediness is low (market fear) and buying when greediness is high (FOMO).

After this process the agent will create an order, hopefully the order will be matched, if not it will remain in the ledger for other 4 iterations, after which it will be eliminated.

6.3.4 Learning

Each Agent will adjust its three decision parameters (Truesight, Trendsight, and Greediness) at the conclusion of each time interval. It does so by determining if its previous activities led to an increase in Net Worth, if that is the case the Agent will most likely keep the parameters unchanged, with lower probability it will repeat the previous choice made (thus if he saw a surge in its net worth and in the previous iteration he decided to increase (decrease), let's say, its Greediness, it will increase (decrease) it again, and with very little probability it will do the opposite of what it did at the previous iteration.

In mathematical terms:

Let

P: payoff of an Agent, it's how much its Net Worth increased since the last time it learned.

$S[y]$ $with$ $y \in [T_s, D_s, G]$: Strategy the Agent is choosing at this time for the parameter y.

$S^{old}[y]$: strategy with which the Agent learned at the previous loop. If $S^{old}[y] = -1$ it means the agent decreased its y-decision parameter, if $S^{old}[y] = 1$ it increased it, if $S^{old}[y] = 0$ it did nothing.

$w_k^{P>0}$: Probabilities to pick a new strategy among $S[y] = 1$, $S[y] = 1$, or $S[y] = 0$ given the previous strategy $S^{old}[y] = k$ and that P > 0.

$w_k^{P\leq 0}$: Probabilities to pick a new strategy among $S[y] = 1$, $S[y] = 1$, or $S[y] = 0$ given the previous strategy $S^{old}[y] = k$ and that P $\leq$ 0.

$\Gamma\left(w_k^P\right)$: Function outputting S^{new} based on the probabilities w_k^P

if $P \leq 0$ **then**
      for y in $[T_s, D_s, G]$ **do**
           if $S^{old}[y] = 1$ **then**
           $S^{new}[y] = \Gamma\left(w_1^{P\leq 0}\right)$
           else if $S^{old}[y] = 0$ **then**
           $S^{new}[y] = \Gamma\left(w_0^{P\leq 0}\right)$
           else if $S^{old}[y] = -1$ **then**
           $S^{new}[y] = \Gamma\left(w_{-1}^{P\leq 0}\right)$
           end if

else if $P > 0$ **then**
      for y in $[T_s, D_s, G]$ **do.**
           if $S^{old}[y] = 1$ **then**
           $S^{new}[y] = \Gamma\left(w_1^{P>0}\right)$
           else if $S^{old}[y] = 0$ **then**
           $S^{new}[y] = \Gamma\left(w_0^{P>0}\right)$
           else if $S^{old}[y] = -1$ **then**
           $S^{new}[y] = \Gamma\left(w_{-1}^{P>0}\right)$
           end if

end if

6.3.5 *Imitation*

The imitation process is relatively simple: the agents are connected in a Barabasi-Albert Graph in which the wealthier nodes are also those with a higher degree. Each tick, agents will examine their first-degree neighbours and adjust their parameters to resemble those of their wealthier neighbours.

Let:

A_i: i-Agent

N^i: Set of the Neighbours of the i-Agent

N^i_j: j-Neighbour Agent of the i-Agent

$W(x)$: Net Worth of the element $\times$ (an Agent or a Neighbour)

$G(x)$: Greediness of the element $\times$ (an Agent or a Neighbour)

$T(x)$: Truesight of the element $\times$ (an Agent or a Neighbour)

$D(x)$: Trendsight of the element $\times$ (an Agent or a Neighbour)

$\deg(x)$: Degree of the node associated to the element $\times$ (an Agent or a Neighbour) in the Network

V: number of Agents in the model

for N^i_j *in* N^i **do**

    if $\left(N^i_j\right) > W(A_i)$ **then**

        if $G\left(N^i_j\right) \geq G(A_i)$ **then**

$$G(A_i) = G(A_i) \cdot \left(1 + \frac{deg\left(N^i_j\right)}{V}\right)$$

        else

$$G(A_i) = \frac{G(A_i)}{\left(1 + \frac{deg\left(N^i_j\right)}{V}\right)}$$

        end if

        if $T\left(N^i_j\right) \geq T(A_i)$ **then**

$$T(A_i) = T(A_i) \cdot \left(1 + \frac{deg\left(N^i_j\right)}{V}\right)$$

        else

$$T(A_i) = \frac{T(A_i)}{\left(1 + \frac{deg\left(N^i_j\right)}{V}\right)}$$

        end if

        if $D\left(N^i_j\right) \geq D(A_i)$ **then**

$$D(A_i) = D(A_i) \cdot \left(1 + \frac{deg\left(N^i_j\right)}{V}\right)$$

        else

$$D(A_i) = \frac{D(A_i)}{\left(1 + \frac{deg\left(N_j^i\right)}{V}\right)}$$

end if

end if

6.3.6 Results and Comments

When looking at the mean value of both decisional parameters (Truesight and Trend-sight), we observed that in certain moments the population prefers to follow a pricing strategy where the fundamental value (truevalue) of a stock has more weight in the decision process than the trend (trendvalue) and vice versa. This is equivalent to say that there is no convergence when it comes to the strategy adopted. However, this phenomenon completely disappears when we inhibit either one among the imita-tion or the learning mechanism. This led us to think that to obtain non trivial model, where we can observe the emergence of types of investors, we need both mechanisms working at the same time.

Other interesting results came from performing a K-Mean Cluster analysis over the time series of the behavioural parameters. We found that there are 3 main distinct behavioural clusters over time, this was the objective of the whole simulation, finding dynamic types of investors. Translated into the real world, it is like to identify groups of investors who react similarly to changes in the market in which they participate by interacting (imitation mechanism) and by changing their strategy based on previous performance (learning mechanism).

6.4 Conclusions

ABM is a powerful and innovative tool which ideal use is to tackle problems hard to face with a top-down approach. Thinking about financial markets as complex systems makes them an obvious area of application for ABM. However, unlike ML, which already shows promising real world applications, ABM for now seems providing most of its results in the academic world, especially to describe social phenomena happening in the markets. This leads us to think that we are still at an early stage for this modelling technique, thus there is plenty of room for improvement and innovation. Our model was on the same line. A very early stage research on investors dynamic clustering and on the role of imitation and learning in the modelling of financial markets.

In conclusion we can say that while ABM is still waiting to escape academia, it's usefulness is undeniable, and the financial world can only benefit from it.

References

J.D. Farmer, P. Patelli, I.I. Zovko, The predictive power of zero intelligence in financial markets, in *Proceedings of the National Academy of Sciences* (2005)

D. Hirshleifer, T.H. Slew , Herd behaviour and cascading in capital markets: a review and synthesis, Eur. Finan. Manage. (2003)

G. Iori, J. Porter, *Agent-based modelling for financial markets* (University of London, City, 2012)

B. LeBaron, R. Yamamoto, Long-memory in an order-driven market. Physica A: Stat. Mechan. Appl.

B. LeBaron, Agent-based computational finance: suggested readings and early research. J. Econ. Dyn. Control **24**, 5–7 (2000)

H. Simon, *Models of Bounded Rationality: Empirically Grounded Economic Reason* (M.I.T. Press, 1997)

Riccardo Vasellini PhD candidate with a background in civil, environmental, and management engineering. I'm currently studying at the University of Siena's Department of Information Science, majoring in the topic of Complex Systems. In the scientific world, my objective is to use Agent Based Modelling and Artificial Intelligence to get significant insights into Complex Systems. I feel that studying Emergent Phenomena and Systems Dynamics is vital for making sound decisions. While studying on my Doctorate, I work as a Project Manager in a variety of industries, including Real Estate Development, Real Estate Portfolio Management, Renewable Energies, and Tourism.

Chapter 7
ML Application to the Financial Market

Riccardo Vasellini

Abstract The globe is being swamped by data, and the rate of fresh data gathering is increasing exponentially. This was not the case only two decades ago, and even if there were technological constraints to the use of machine learning, the lack of data to feed the algorithms constituted an additional obstacle. Furthermore, if acquiring precise and meaningful data results in too high a cost, it may be more cost-effective to acquire data that is not directly related to the financial phenomena we need to analyze, a so-called alternative dataset. The ultimate purpose of alternative data is to provide traders with an informational advantage in their search for trading signals that yield alpha, or good investment returns that are unrelated to anything else. A strategy may be based purely on freely available data from search engines, which ML systems could then correlate to some financial occurrence.

Keywords AI · Machine learning

7.1 Introduction

The last decade has witnessed a significant increase in data storage and collection in numerous industries, including the financial sector. This enabled financial institutions, investors, insurance companies, and anybody with an interest in the movement of stocks, derivatives, and interest rates to use Machine Learning techniques that previously lacked the computational capacity and data to be utilized effectively.

AI and ML can be applied to a wide variety of financial sectors, including portfolio strategy formulation, fraud and unlawful activity detection, risk assessment and management, legal compliance, algorithmic trading, etc.

In this chapter we will presents examples of applications found in literature., modelling techniques and paradigms of ML in finance. We will conclude assessing the limits and dangers of using these techniques and what the future ahead looks like.

R. Vasellini (✉)
Universitá di Siena, Siena, Italy
e-mail: r.vasellini@student.unisi.it

© The Author(s), under exclusive license to Springer Nature Switzerland AG 2023
F. Cecconi (ed.), *AI in the Financial Markets*, Computational Social Sciences,
https://doi.org/10.1007/978-3-031-26518-1_7

7.2   Fundamentals

Machine Learning can be grouped in three main area: Supervised, Unsupervised and Reinforcement Learning.

Supervised learning is probably the simplest form of ML, linear regression is an example of one of its algorithms and it has been used for decades before the terms "AI" or "ML" even appeared. Other than linear regression, supervised learning comprehends also other algorithms, such as Logistic regression, Decision Trees, K-Nearest neighbour, Support Vector Machines, Neural Networks. The main characteristic is that of using a dataset containing labels associated to the features of the data points. The whole point is to find a function able to map the data points features to their respective labels after training the algorithm to do so with part of the labelled data. This paradigm assumes that we have knowledge of the answers we are looking for. Once trained and tested the map found by the algorithm can be used on unlabelled data. This can be done either for classification or regression problems.

Unsupervised learning involves recognising patterns within data. It usually entails searching for clusters, or groupings of similar findings. ML algorithms, in practice, associate similar things together, where the similarity is often expressed in terms of distance from a cluster centre. A good algorithm is agnostic to the scale of the data, therefore it's common to employ feature scaling techniques such as z-score normalization or min–max methods. The most used clustering technique is the k-means clustering, however other techniques such as distribution based, density based or hierarchical clustering are used.

Reinforcement learning revolves around multi step decision making. An agent takes actions based its estimate of rewards and costs taken in the environment he is in, which is described by a series of states, typically as a Markov Decision Process. The reward (net of costs) is calculated using the Q-function, and a particular action is the best if it lead to a state with the highest Q-value. An important concept is that of exploration, an "intelligent agent" has always the possibility to perform a random action, this is done in order to not get stuck into exploiting a state that provides rewards without exploring different states that might lead to better outcomes. The agent doesn't need labelled data.

What is fuelling the adoption of ML is not only the access to these new techniques presented. The world is being submerged by data, and the rate of acquisition of new data grows exponentially. This was not the case only two decades ago, and even if there were technological restrictions to the application of machine learning, the lack of data to feed the algorithms imposed an additional barrier. Moreover, when gathering accurate and valuable data results too expensive, it might be more cost-effective to acquire data that is not directly connected to the financial phenomena we need to analyse, a so-called alternative dataset. The ultimate goal of alternative data is to give traders an informational edge in their search for trading signals that produce alpha, or positive investment returns that aren't linked to anything else (Jensen 2018). A strategy might be entirely built on freely available data from search engines, which might be correlated to some financial phenomenon by ML algorithms. As an example, think of the hedge fund Cumulus, which simply used weather data to trade farming companies.

Finally, by combining ML algorithms, contemporary computer power, and a massive quantity of data It is possible to construct strong applications for the financial sector, which we will now explore.

7.3 Applications

7.3.1 Portfolio Management

Since being presented by Markowitz in 1952, the mean–variance optimization (MVO) has been the main paradigm upon which build portfolios. MVO success is strictly linked to that of quadratic programming (QP), that makes solving MVO straightforward (Jensen 2018). One alternative proposed is that of the risk budgeting approach, where, put simply, the portfolio has risk budgets for each asset type and we try to allocate assets risks according to these budgets. This approach dominates MVO in various field of PO. Unfortunately, it usually involves dealing with non-linearities making it hard to solve. However, in the same way ML evolved in the latest years, also PO evolved, while QP was chosen for its easy computational approach nowadays a whole new set of algorithms, more computationally expensive, are used. Among these we find algorithms developed for large scale ML problems: coordinate descent, alternating direction method of multipliers, proximal gradient and Dykstra's algorithm. Using these algorithms allows for surpassing the QP paradigm and delve into a whole new set of models not limited by linearity. The future of PO is to use these ML algorithms to develop new PO models, or simply use already developed ones which are now computationally more approachable then in the past.

As an example, for this approach, Ban, El Karoui and Lim (Perrin 2019) proposed in 2018 a performance-based regularization (PBR) and performance based cross-validation models to solve portfolios optimizations problems in order to go beyond the estimation issues resulting from applying classical methods to real data. Regularization is a technique used for decades to solve issues revolving around problems set as linear but for which, in reality, small variations of the constants of the problem lead to big deviations in the solution. In practice is a type of regression that constrains or reduces the estimated coefficients to zero. Thus, it prevents the risk of overfitting, by discouraging the learning of a more complicated model. In practice reduces the variance of the model without increasing too much its bias.

The authors regularize a portfolio optimization problem with the purpose of improving the out-of-sample performance of the solution. To achieve this, they constrain the sample variances of the portfolio risk and mean. The goal is to create a model which find a solution to the portfolio problem (whether it is the traditional problem or the CVaR one) with very low bias and high out-of-sample performance.

On another paper (Jensen 2018), Perrin and Roncalli, individuate what can be considered the four most important algorithms for portfolio optimizations: coordinate descent, alternating direction method of multipliers, the proximal gradient method and the Dijkstra's algorithm. The authors assess that the success of the MVO paradigm lies in the absence of competing implementable models. The reasons individuated are hard to estimate parameters and complex objective functions, using a combination of the four-algorithm mentioned allows for framework able to consider allocation models outside of the QP form.

Finally, an innovative approach is presented by Ban et al. (2018). The authors try to tackle the problem of data heterogeneity and environmental uncertainty in portfolio management. They do so using Reinforcement Learning. Specifically, to include heterogeneous data and improve resilience against environmental uncertainty, their model (State-Augmented RL, SARL) augments asset information with price movement predictions, which may be exclusively based on financial data or obtained from nontraditional sources such as news. Tested against historical data for Bitcoin prices and High-Tech stock market they validate this method showing simulated results for both total and risk adjusted profits.

In general, ML paired with computing power advances, is allowing scientists and professional to test model less bounded by quantitative restrictions, solutions once impractical are becoming within reach.

7.3.2   Risk Management

Risk Management (RM) has seen an increase in the adoption of both new and old models that deal with a large number of variables and data thanks to the use of ML algorithms which now can find answers in an acceptable amount of time.

Risk permeates the world of finance and beyond. We will examine how machine learning is used to some of its most prevalent declinations.

When dealing with credit risk knowing the probability that a debtor will repay is crucial knowledge for financial institutes (or a lender in general). This is particularly hard when dealing with SME or retail investors for which the data available are sparse and sometimes inaccurate. The general idea of the various models that can be found in literature is that using ML is possible to find patterns in the behavior of these small borrowers using data which are not traditionally linked to predicting credit risk (Ban et al. 2018).

Another area where ML is thriving, is assessing the credit risk of a complex derivate object such as credit default swap (CDS). For these objects a deep learning approach has shown better results than traditional, in their research Son, Byun and Lee models (Ye et al. 2020) showed the parametric models they used consistently had better prediction performance than the benchmark models and, among all the models used, ANN showed the best results.

Currently, the primary use of machine learning for managing risk when trading in markets is the validation of proposed models via back testing on vast amounts of data. Another key application is understanding how trading will affect an illiquid market, hence altering the price of the traded asset. The difficulty of high-volume trading of a single asset can be circumvented by employing machine learning algorithms to identify similar assets.

An interesting application within this framework is from Chandrinos, Sakkas and Lagaros (Lynn et al. 2019). These researches developed a tool that uses ML as a risk management tool for investments. Specifically, the investigation is centred on the categorization of the signals generated by a trading strategy into those that are successful and those that are not profitable through the use of artificial neural networks (ANN) and decision trees (DT). To do this they use two previously proposed currency portfolios and using their Artificial Intelligent Risk Management System (AIRMS) improve their performance by reducing the losses rather then increasing the gains. In their back tests not only did the two methods employed (DT and ANN) boosted the profitability of the portfolios, but they also significantly improved their sharpe ratio by decreasing their standard deviation.

Another application is that of Sirignano, Sadhwani and Giesecke (Son and Lee 2016), who used a deep learning model to analyse data of over 120 millions mortgages issued in the US between 1995 and 2014. Trying to understand the probabilities of certain borrowers behaviour and the risk of incurring in a non performing loan, they looked at various variables, financial as well as macroeconomic. They were able to conclude that one of the most relevant factors in predicting the success of a mortgage is the unemployment rate of the zip code in which the mortgage was issued, highlighting the link between housing finance markets and macroeconomy.

7.3.3 PropTech

PropTech is the widespread adoption of emerging technologies within the real estate industry. Such technologies include home matching tools, drones, virtual reality, building information modelling (BIM), data analytics tools, artificial intelligence (AI), Internet of Things (IoT) and blockchain, smart contracts, crowdfunding in the real estate industry, financial technologies (fintechs) related to real estate, smart cities and regions, smart homes, and the shared economy (Spyros et al. 2018). The real estate industry goes hand in hand with the financial market especially when it comes to listed Real Estate Investments Trusts (REIT) which in the US have more than 1 trillion USD market capitalization (Sirignano et al. 2018).

In this context being able to accurately predict home prices has clear relevance. Unfortunately to do so accurately requires deep knowledge of the local market and of the surroundings, a type of knowledge which is impractical to obtain for big REITs operating with great amount of properties or credit institution pricing hundreds of collaterals every day.

A solution is proposed by Siniak et al. (2020). The authors argues that commonly used indexes scarcely accurately depict real estate markets at a granular level. In fact, they try, using over 16 years of home sale data, to arrive at a pricing prediction accuracy of a single house.

The method proposed is called gradient boosted home price index, which uses the ML technique of gradient boosted regression trees algorithm, which constructs multiple decision trees and recursively fit a model. In practice they reverse the multiple decision tree building process, which would lead to over fitted classification trees (one leaf for each house), building a low complexity decision tree (weak learner), then building other low complexity trees where the splits are performed when poor predictions happen. The final tree obtained is called a strong learner and can be thought as the weighted average of the weak learners. The authors argue that this method is able to predict home prices at a singular level in a better way than the classical indexes available, which would be of great use for financial institutions and real estate funds.

7.3.4　Asset Return Prediction

Predicting the exact right price of an asset is the holy grail of finance. Being able to do so would translate in riches and wealth, therefore asking how to do it with a small as possible error is certainly a non-trivial question. In fact, one of the most important results of financial sciences, the Black–Scholes model, is used for exactly this reason: pricing a form of asset, specifically, an option (Caporin et al. 2021).

Asset prices follow a highly non-linear behaviour, suffer from feedback loops and, sometimes, boom and bust cycles (Barr et al. 2016), a powerful tool to deal with these features is certainly deep learning. The problem of predicting the price of an asset is equivalent to that of predicting the behaviour of a time series and in economics and finance this is often done using Dynamic Factor Models (DFM). DFM can be thought as models of the co-movement of multiple timeseries.

Deep learning can be used to integrate these models. As an example, we can look at Feng, He and Polson (Black and Scholes 1973). To anticipate assets returns, the authors developed dynamic factor models trained using deep learning. Using stochastic gradient descent, both hidden components and regression coefficients are computed simultaneously, thus leading to increased out-of-sample performance compared to traditional models.

7.3.5　Algorithmic Trading

Algorithmic Trading consist in the automation, through a computer, of all, or part of, the steps needed to execute a certain trading strategy. It can be seen as a fundamental part of quantitative trading, and, according to Wall Street data, Algorithmic trading accounts for around 60–73% of the overall US equity trading. (Feng et al. 1804).

The way quantitative trading has evolved can be synthetized into three main phases:

In the first era ('80s–'90s), quantitative firms would use signals derived from academic research, often using a single or very few inputs coming from the market or fundamental data. Strategy where then pretty simple, the difficulty lied in obtaining the right data quickly.

In the second phase (2000s), to explore arbitrage opportunities, funds employed algorithms to identify assets vulnerable to risk variables such as value or momentum. In this phase factor-based investing was mostly employed, and it's the factor-based industry to have caused the quant quake of August 2007.

The last phase it's the one we are living, in which funds are using investments in machine learning and alternative data to develop effective trading signals for recurring trading methods. In this extremely competitive environment, once a valuable anomaly is uncovered, it swiftly disappears owing to competition. (Intelligence 2022).

The primary objective of using machine learning to trading is to forecast asset fundamentals, price movements, or market circumstances. A strategy may use numerous machine learning (ML) algorithms to achieve this. By incorporating forecasts about the prospects of individual assets, capital market expectations, and the connection across securities, downstream models may provide portfolio-level signals.

Many trading algorithms use technical indicators from the markets to adjust a portfolio composition increasing it's expected return or reducing its risk.

As application example we can look at the use of Deep Learning by Lei, Peng and Shen to improve a commonly used indicator in technical analysis, the Moving Average Convergence/Divergence (MACD) (Lei et al. 2020). The authors start by noting how classical MACD techniques fail to understand the magnitude of trend changes, this can lead to signal the algorithm to trade when actually there are no big trend changes on the horizon but just fluctuation. This would lead to unnecessary losses due to transaction costs. Therefore, thanks to Residual Networks, is possible to estimate certain characteristics of the time series representing the stock traded, in particular the authors focus on the local kurtosis of the time series where the MACD indicators would signal to trade. If the Residual Network estimates a kurtosis higher than 0 (meaning a higher steepness of the curve than a normal distribution), the algorithm will trust the trading point indicated and perform a trade, otherwise it will ignore it and hold the position.

Tested on the CSI300 stock index the algorithm proposed outperformed the classic one employing only the MACD, showing how existing algorithmic trading strategies can be improved by ML.

7.4   Risks and Challenges

While providing many advantages, the use of ML in Finance is not immune to its common pitfalls.

In all fields, including finance, the complexity of a machine learning model is the primary factor contributing to its riskiness. The algorithms for machine learning are fundamentally quite difficult since they operate on massive and sometimes unstructured data, such as texts, photos, and sounds. As a result, training of such algorithms requires a complex computational infrastructure as well as a high degree of expertise and understanding on the side of the modellers (Sen et al. 2021). Moreover, complexity might make the algorithms hard to be implemented by final users, which could have computing infrastructures not apt to render an answer within the time constrains imposed by the task.

Most of ML models are *black boxes,* this means that the user might know what the answer to the posed problem is, but have no idea of how it was found. This can lead to several problems and legal disputes. Consider, as an example, an insurance firm that relies on ML to assess the premium of a contract. Without the ability to explain how the insurer estimated the value, the client may be hesitant to sign the contract with the underwriter. Similarly, if a bank denies a customer access to credit without providing a reason, the bank may face consequences. While many applications (think of suggesting a product to a customer) do not need model interpretability, many others do, and the inability to do so may hinder the progress in applying ML to the financial world.

Another challenge is presented by biases. The way data is collected might affect the answers provided by the ML algorithms employed. The readily available data is intrinsically biased towards replicating existing practices. Think of data coming from credit institutions where customers are classified based on the type of loan they received (or not received). Using ML, we might simply keep replicating the same decision process used in the past, without innovating anything and, in fact, rediscovering the wheel. Biases can emerge for gender or race, creating an ethical problem and exposing the user to discrimination lawsuits. To avoid such biases is essential to perform accurate data cleaning, feature selection and extraction.

Ulterior risks arise from adversarial ML, which is the practice of feeding erroneous data to a ML algorithm in order to fool it towards certain results. Famous is the case of Navinder Sarao (Wang 2015), a young Briton who contributed to trigger the 2010 flash market crash by creating lots of orders and then cancelling them (a practice known as spoofing), thus inducing algorithmic trading bots to modify their strategies and artificially modify the market. The best solution against adversarial ML is human monitoring, however, when the algorithm is a black box, this becomes challenging and more interpretability is required in order to be successful.

Finally, another major concern is certainly how to not breach data privacy, this is especially true when it comes to sensible financial information. Many governments are making steps forward in tackling the issues coming from handling personal data, the most notable action is arguably the EU GDPR which became effective in May 2018.

7.5 Conclusions

Nowadays, ML permeates every industry in a way or another. Finance has both, an industry and a research field, has always been inclined to computational approaches and using ML can only be seen as a natural progression of this attitude.

We explored various applications, notably most of them seems to be revolving around portfolio management and optimization. In fact, even when we are talking about risk management we might as well be talking about a sub-field of portfolio optimization, while algorithmic trading can be seen as an automatic implementation of portfolio management rules and models.

Other important fields are those of pricing assets and categorizing customers. What seems to be crucial is not only the type of model used, but the cleanness of the data acquired. Using the right data is fundamental to avoid biases and to obtain innovative insights. However, it is a delicate process since it's easy to breach personal privacy when handling financial data.

Major concerns come from the fragility of complex models, which often are black boxes to their users whom, not understanding the model used, might be prone to be exploited, an example of it is the practice of "spoofing".

In conclusion, ML is declining a great wave of innovation into the financial world, and as always with big industry changes, monitoring and regulating is essential to avoid speculations and exploitations.

References

G.-Y. Ban, N. El Karoui, A.E.B. Lim, Machine learning and portfolio optimization. Manage. Sci. (2018)

J.R. Barr, E.A. Ellis, A. Kassab, C.L. Redfearn, N.N. Srinivasan, B.K. Voris, Home price index: a machine learning methodology. Encycl. Seman. Comput. (2016)

F. Black, M. Scholes, The pricing of options and corporate liabilities. J. Polit. Econ. **81**(3) (1973)

M. Buchanan, *Forecast: What Physics, Meteorology, and the Natural Sciences Can Teach Us About Economics* (Bloomsbury, 2013)

M. Caporin, R. Gupta, F. Ravazzolo, Contagion between real estate and financial markets: a Bayesian. North Am. J. Econ. Finan. (2021)

G. Feng, H. Jingyu, N.G. Polson, *Deep Learning for Predicting Asset Returns* (2018) arXiv:1804. 09314v2

S. Jensen, *Hand-On Machine Learning for Algorithmic Trading* (2018)

Y. Lei, Q. Peng, Y. Shen, Deep learning for algorithmic trading: enhancing MACD, in *ICCAI '20: Proceedings of the 2020 6th International Conference on Computing and Artificial Intelligence* (2020)

T. Lynn, S. Aziz, M. Dowling, *Disrupting Finance* (2019)

Mordor Intelligence, *Algorithmic Trading Market—Growth, Trends, Covid-19 Impact, and Forecasts (2022–2027)* (2022)

S.A.R.T. Perrin, *Machine Learning Optimization Algorithms & Portfolio Allocation* (SSRN, 2019). https://ssrn.com/abstract=3425827 or https://doi.org/10.2139/ssrn.3425827

J. Sen, R. Sen, A. Dutta, *Machine Learning: Algorithms, Models and Applications* (2021)

N. Siniak, T. Kauko, S. Shavrov, N. Marina, *The Impact of Proptech on Real Estate Industry Growth* (2020)

J.A. Sirignano, A. Sadhwani, K. Giesecke, *Deep Learning for Mortgage Risk* (2018)

Y. Son, H. Byun, J. Lee, Nonparametric machine learning models for predicting the credit default swaps: an empirical study. Exp. Syst. Appl. (2016)

C.K. Spyros, G. Sakkas, N.D. Lagaros, AIRMS: a risk management tool using machine learning. Exp. Syst. Appl. (2018)

Y.-Y. Wang, Strategic spoofing order trading by different types of investors in the futures markets. Wall Strett J. (2015)

W. Wang, N. Yu, A machine learning framework for algorithmic trading with virtual bids in electricity markets. EEE Power Ener. Soc. Gener. Meeting (2019)

Y. Ye, H. Pei, B. Wang, P.-Y. Chen, Y. Zhu, J. Xiao, B. Li, Reinforcement-learning based portfolio management with augmented asset movement prediction states, in *The Thirty-Fourth AAAI Conference on Artificial Intelligence (AAAI-20)*

Riccardo Vasellini PhD candidate with a background in civil, environmental, and management engineering. I'm currently studying at the University of Siena's Department of Information Science, majoring in the topic of Complex Systems. In the scientific world, my objective is to use Agent Based Modelling and Artificial Intelligence to get significant insights into Complex Systems. I feel that studying Emergent Phenomena and Systems Dynamics is vital for making sound decisions. While studying on my Doctorate, I work as a Project Manager in a variety of industries, including Real Estate Development, Real Estate Portfolio Management, Renewable Energies, and Tourism.

Chapter 8
AI Tools for Pricing of Distressed Asset UTP and NPL Loan Portfolios

Alessandro Barazzetti⊙ **and Rosanna Pilla**

Abstract The Non- Performing Exposures are an asset financial class that sees large volumes of credits of banking origin traded on the free market through assignment operations. The issue of credit enhancement therefore acquires a central role in the credit purchase process. In this study we describe an approach to the issue of valuing Non Performing Exposures based on AI algorithms and calculation methods rather than traditional econometrics. The following approach has been studied for the Italian NPE market, as it has unique characteristics in terms of size, legal and technical-real estate aspects.

Keywords NPE · Loan evaluation · AI · Fintech · Business plan · Cluster tree analysis · Expert system · Cash flow

8.1 Non-performing Loans

NPE (Regulation (EU) 2013) refers to a credit position of banking origin, for which the debtor has no longer been able to honor payments. The debtor enters a state of insolvency and the credit is classified as non-performing becoming a Non Performing Exposure or NPE. Three different types of NPE are identified based on the number of months that have passed since the last payment made by the debtor to the institution that granted the credit.

Each type corresponds to a precise accounting fulfillment and calculation of interest and expenses up to the judicial procedure for the compulsory recovery of the credit through the judicial auction mechanism.

A. Barazzetti (✉)
QBT Sagl, Via E. Bossi 4, 6830 Chiasso, Switzerland
e-mail: alessandro.barazzetti@qbt.ch

R. Pilla
Reacorp Sagl, Via E. Bossi 4, 6830 Chiasso, CH, Switzerland
e-mail: rosanna.pilla@reacorp.ch

The types of NPE are identified as follows, in order of time since the last payment made:

1. Overdrawn and/or past-due exposures
2. Unlikely-to-pay exposures (UTP)
3. Non-performing loans (Bad loans).

The classification of NPE positions was defined by the European Union through a regulation issued by the European Banking Authority (EBA) (European Central Bank 2017) pursuant to a European Regulation (EBA/GL 2016).

The NPE positions are also distinguished by the type of guarantee linked to the credit: therefore we have non-performing positions of loans with a mortgage guarantee (Secured loans) and loans without a mortgage guarantee (Unsecured).

Secured positions, the mortgage guarantee is made up of the debtor's real estate assets: therefore, these are non-performing loans whose recovery is facilitated by the possibility of proceeding through the courts at the property auction.

Unsecured loans, on the other hand, are distinguished on the basis of the type of credit disbursed: we have credits of a financial, commercial and banking nature.

Another important distinction concerns the type of debtor. The credits can be disbursed to individuals or legal entities. The type of debtor has an impact above all in the recovery legal action: for legal persons, the judicial recovery of the amounts due takes place through a bankruptcy executive procedure (ie bankruptcy) compared to individual execution (ie attachment) for natural persons.

Furthermore, the credit amounts for legal entities are usually higher than for natural persons.

The study will focus on the Italian NPE credit market since it is a large market as well as being particularly complex in relation to the legal credit recovery activities and the particularity of the properties placed as collateral for the credit.

8.2 The Size of the NPE Market

The market for bank credit positions classified as NPE has a considerable size in terms of both overall volumes and number of credit positions.

At the end of 2021, the stock of NPEs in Italy amounted to €330 billion, of which €250 billion in NPL positions and €80 billion in positions classified as UTP (Market Watch 2021).

is expected in consideration of the current economic crisis, with strong repercussions in the three-year period 2023–2026 in relation to the high rate of inflation, which is affecting the real economy of the country.

Of the figures previously indicated, 75% regards loans to companies and 70% loans with real estate guarantees (secured).

NPE transactions in 2021 amounted to approximately 36 billion euros for 34 billion euros of NPLS and 2 billion euros of UTP divided as follows: 80% of loans to companies and 50% of loans secured by real estate (Fig. 8.1).

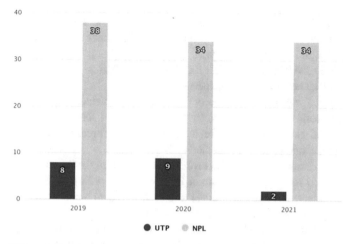

Fig. 8.1 NPE portfolio sales in the last three years in billions of Euros

8.3 The Valuation of Credits and Credit Portfolios

Non-performing loans or NPEs are purchased by investment funds through Special Purposes Vehicle (SPV) through the credit securitization process (Bank of Italy 2017).

The purchase of credits takes place through a non-recourse assignment by the assignor bank according to a price that is negotiated between the parties at the time of purchase.

An alternative to the non-recourse transfer is represented by the Contribution Funds or FIAs, which allow the banking institutions that transfer the credits to transfer the NPE positions in a closed fund. In exchange, the assignors receive the units of the fund in proportion to the amount of impaired loans transferred. AIFs are not speculative in nature.

In both cases, banks can assign or contribute single positions or single names to the AIFs or entire loan portfolios which also group together a large number of debtors.

The purchase price of the credit or of the specular contribution value for the AIFs is determined by estimating a credit recovery forecast, known as the Business Plan of the transfer transaction.

The activity that allows the buyer to arrive at the determination of the recovery forecast and therefore the drafting of the Business Plan takes the name of credit due diligence and consists in the analysis of the single NPE positions.

The set of information collected by the buyer and made available by the seller is then used by the econometric models to elaborate the recovery forecast or Business Plan.

Usually the buyer relies on companies specialized in technical-legal advisory to carry out the due credit due diligence: these companies have staff who are experts in legal, financial and real estate analyses.

In the case of AIFs, on the other hand, the contributing institutions rely on the figure of the Independent Credit Expert.

The purchase price of receivables or the contribution value, whether single names or entire portfolios of receivables, is determined by discounting the cash flows included in the Business plan, which represents the forecast of credit recovery, according to the rate of return expected or IRR.

The rate of return represents the investor's expectation for the sale transaction.

The expected return rates or market IRRs differ according to the type of credit but we can indicate a range between 11 and 15% as a benchmark for Secured credit portfolios, with a highly speculative investor approach.

Of particular importance is the definition of the rate of return for FIA contribution operations, which we remind you do not have a speculative purpose.

For this type of transaction, the expected rate of return is set at ~ 3%: the rate is defined on the basis of a series of objective market parameters linked to the ten-year average of the Euribor and the average of the savings deposit rates of the main banking operators.

The econometric methods for determining the purchase price of credit or credit portfolios are essentially statistical in nature based on the analysis of the correlations between the descriptive attributes of the debtor and the nature of the credit in relation to the historical trend of collections.

Alternatively, we have implemented a method for determining recovery forecasts for UTP credits based on an AI approach.

8.4   Description of an AI Valuation Method of UTP Credits

In this chapter we describe a credit valorisation method based on an AI approach based on the construction of an Expert System (Bazzocchi 1988).

With Expert System we mean a program able to replicate the behavior of one or more expert people (Iacono 1991).

The founding principle of the Expert Systems is the writing of "rules" which constitute the basis of the knowledge of the program: these rules are built thanks to interviews with experts in the sector, thanks to statistical analysis of data to extrapolate regularities identified by means of Decision Trees[1] or using a Machine Learning solution (Woolery and Grzymala-Busse 1994).

We focus our study on UTP Secured and Unsecured credits, disbursed to both individuals and legal entities.

UTP credits are defined as credits that have not yet become non-performing: for this type of credit, the debtor has stopped paying his debt but the bank has not yet

---

[1] V.A. Manganaro, DM Techniques: Decision Trees and Classification Algorithms (Palermo, Italy).

classified it as actual insolvency. The UTP condition has a legally limited duration over time and has two potential exit conditions: the position becomes a real non-performing loan or NPL or, through a specific recovery strategy, it returns to the performing condition.

Both exit conditions determine a credit recovery forecast and a related recovery time frame.

Our program or algorithm for determining the valuation of UTP credits includes the following flow chart, which we will analyze in detail:

The UTP position is initially distinguished between a natural person and a legal person.

If the position is a natural person the information that is used is of a commercial nature.

If the position is instead a legal person, the information, in addition to the legal ones, concerns the company's balance sheet data and sector analysis.

Here is the information used in the model, which represents the input of the program (Tables 8.1 and 8.2).

Table 8.1 List of variables used in the algorithm for individuals

Input parameters	Input type	Description
GBV	Amount	A Due credit
How much did he pay	Amount	B Credits already paid
How much disbursed	Amount	C Credit disbursed
When last payment	At your place	
Property location	String	
Real estate value	Amount	
Other Warranties	Yes No	
Guarantors	Yes No	
Average installation amount	Amount	
Collections from utp	Yes No	Collections after switching to UTP
Work	Yes No	
Salary	Amount	F amount of salary
Work type	Drop down menu	
Profession	Drop down menu	
Marital status	Drop down menu	
Bride works	Yes No	
Sons	Yes No	
Available	Yes No	
Special	Yes No	

Table 8.2 List of variables used in the algorithm for legal entities

Input parameters	Input type	Description
GBV	Amount	A Due credit
How much did he pay	Amount	B Credits already paid
How much disbursed	Amount	C Credit disbursed
When last payment	At your place	
Property location	String	
Real estate value	Amount	d
Other Warranties	Yes No	
Guarantors	Yes No	
Average installation amount	Amount	AND
Collections from utp	Yes No	
Sector	Drop down menu	Merchandise sector of the company
active	Amount	Budget items
Passive	Amount	Budget items
Members	Yes No	
Activate	Yes No	
Refundable	Yes No	
Special	Yes No	
Refinancing analysis	Drop down menu	Index from 1 to 10

Following the scheme in Fig. 8.2, the first analysis consists in determining whether the position remains UTP or will become NPL.

Using the Input parameters, the two UTP credit portfolios were analyzed as follows, through a Cluster Tree Analysis (Table 8.3).

The result of the analysis made it possible to assign the correct weight to each parameter in determining the probability of switching from UTP to NPL.

In the event that the position becomes an NPL, the valuation becomes judicial and involves the application of a haircut on the value of the receivable equal to 80% of the value with a recovery time quantified as an average of 5.3 years (Study of the Times of the Courts Ed. 2022).

In the event that the position remains UTP, it becomes necessary to define what is the most probable exit strategy from the UTP condition towards a return to performing or in any case an out-of-court settlement.

In this case, to determine the recovery forecast, it is necessary to calculate the most probable strategy of non-judicial UTP solutions which can be classified as follows (Table 8.4).

For the determination of the recovery strategies of Table 8.5, for the positions that remain UTP we used a set of rules inserted in the calculation program of the credit recovery forecast.

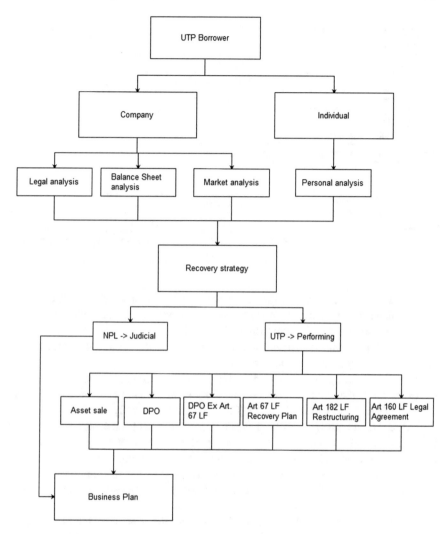

Fig. 8.2 Flowchart

Table 8.3 Portfolios used to analyze NPL vs UTP positions

Portfolio ID	GBV € (~)	Borrower number (~)
Portfolio A	12,000,000.00	30
Portfolio B	55,000,000.00	25

Table 8.4 UTP exit strategies for legal entities

Exit strategy	Description
Asset sale	Consensual asset sale
DPO	Voluntary disposal even with a post or gradual return plan
DPO Ex Article 67	Recovery plan with a DPO
Article 67 lf—recovery plan	Recovery plans
Article 182 bis. restructuring	Legal restructuring plan
Article 160 lf—legal agreement	Legal agreement

Table 8.5 Strategies for exiting the UTP condition for individuals

Exit strategy	Description
Asset sale	Consensual asset sale
DPO	Voluntary disposal even with a DPO or gradual return plan

The Expert System was defined by a group of credit specialists with different backgrounds: we created a team made up of two lawyers, an economist expert in budget analysis and a real estate technical expert.

The working group formulated the rules that define the activation conditions of the various strategies in relation to the input variables.

Particular mention should be made of the parameter relating to sector analysis. By sector we mean the belonging of the company to a precise product category defined through the ATECO codes (ATECO 2007). Through the analysis of sector studies (ISTAT 2021), the group of experts assigned a positive score to each product group in the case of sectors capable of generating GDP, differently a negative score for sectors with recession prospects in relation to macroeconomic conditions current.

In the same way, a company financeability scale linked to specific rules deduced from the analysis of the financial statements of the companies belonging to the portfolios in Table 8.3 was defined.

The analysis of financial statements for legal entities represents the most important element in drafting the rules for determining the most probable credit recovery strategy.

The rules thus determined apply to the parameters entered as input and take the following form (Table 8.6).

Each strategy is then associated with a recovery forecast based on rules (Table 8.7).

The percentages indicated as DiscountN were determined through the statistical analysis of the credit portfolio database in Table 8.3, which have already generated receipts over the years.

The result is highlighted in Table 8.8.

Table 8.6 Example of rules

Rules	Strategy
If FIRST MORTGAGE (SUBSTANTIAL) = YES AND D > A	Consensual asset sale
IF (FIRST MORTGAGE (SUBSTANTIAL) = NO (OR NULL) OR UNSECURED = YES) AND ACTIVE > PASSIVE AND ACTIVATE = YES	Voluntary disposal even with a post or gradual return plan
IF ACTIVE > (PASSIVE + Tolerance * PASSIVE) AND D > 0	Consensual asset sale
IF ACTIVE > (PASSIVE + Tolerance * PASSIVE)	Voluntary disposal even with a post or gradual return plan
IF ACTIVE < = A AND ACTIVE > PASSIVE	Dpo ex art 67 lf recovery plan
IF PASSIVE > ACTIVE AND (A < PASSIVE + Tolerance * PASSIVE AND A > PASSIVE—Tolerance * PASSIVE)	Art 160 lf—legal agreement
…	…

Table 8.7 Enhancement of the strategy for legal entities

Exit strategy	UTP recovery amount
Asset sale	Amount = (min A; D) * % Discount 1
DPO	Amount = A * % Discount 2
…	…

Table 8.8 Recovery values in percentages and in years associated with individual strategies

Strategy	Tolerance (%)	Discount as percentage (%)	Years for recovery
Consensual asset sale	10	90	3
Voluntary disposal even with a post or gradual return plan		80	5
Dpo ex art 67 lf recovery plan	10	80	4
Article 67 lf—recovery plan	10	70	3
Article 182 bis.—legal restructuring plan	20	50	5
Art 160 lf—legal agreement	30	60	4

The credit valorisation thus elaborated makes it possible to generate a business plan of the credit portfolio, assigning to each position:

- Attribution of transition to NPL or permanence in UTP
- If NPL then Strategy is Judicial
- If UTP then the rules are applied to determine the Most Probable Recovery Strategy
- For the selected Strategy, the recovery forecast is determined based on the valuation rules with the application of the Discount percentages calculated statistically as well as the recovery time in years.

The result is a business plan as shown in Fig. 8.3.

Each line of the document represents a debt position for which the credit value or GBV is presented, the legal nature, the exit condition as NPL or UTP and the recovery strategy with the related forecast of cash flows over time and the price of purchase or contribution value calculated as discounting of the cash flow at the reference rate.

8.5 Model Verification

The results of the credit portfolio valuation processing carried out with the program based on the algorithm described were validated by means of a valuation test conducted in parallel and manually with the specialists of the group of experts on the same credit portfolios. This activity has made it possible to consolidate the valorisation rules.

Subsequently, back tests were carried out on the valuation of loan portfolios whose strategy and recovery of the single position was already known. The degree of reliability both in terms of determination of the NPL or UTP conditions and of the exit strategies from the UTP condition has proved to be extremely high as shown in Table 8.9.

8.6 Conclusion

The use of a credit enhancement system based on the use of an Expert System has the advantage of explaining the physical phenomenon underlying the credit recovery process in a transparent and verifiable way. Rules can be subjected to empirical testing over time to check adherence to the physical reality they describe. Furthermore, new rules can be added over time as environmental conditions change. In particular, new regulatory obligations or considerations on the application of macroeconomic parameters.

The application of the model on real credit portfolios allowed us to test the method and verify its correctness on various transactions, with satisfactory results.

NDG	Descrizione debitore	Valore dell'impiego	Natura giuridica	Exit Strategy	Descrizione Strategia	Totale NETTO	31/03/2022	30/06/2022	30/09/2022	31/12/2022
1	Azienda 1	1.417.721,73	PG	UTP	PIANO DI AMMORTAMENTO	1.416.785,53	0,00	18.096,25	18.020,00	24.128,64 **
2	Azienda 2	14.030.445,02	PG	UTP	PIANO DI AMMORTAMENTO	13.890.558,96	0,00	203.596,50	204.326,00	231.908,04
3	Azienda 3	417.389,51	PG	UTP	ART. 160 LF - CONCORDATO PREVENTIVO	247.316,40	0,00	0,00	0,00	0,00
4	Azienda 4	534.772,87	PG	UTP	ART. 160 LF - CONCORDATO PREVENTIVO	16.043,19	0,00	0,00	0,00	8.021,59
5	Azienda 5	126.616,71	PG	UTP	ART. 160 LF - CONCORDATO PREVENTIVO	126.616,63	0,00	0,00	0,00	10.117,29
6	Azienda 6	170.078,15	PG	NPL	DPO CON RIENTRO GRADUALE	17.007,82	0,00	0,00	10.117,29	0,00
7	Azienda 7	688.793,43	PG	NPL	GIUDIZIALE	654.353,76	0,00	0,00	0,00	0,00
8	Azienda 8	1.713.817,88	PG	UTP	DISMISSIONE VOLONTARIA DELL'ASSET	1.628.126,99	0,00	0,00	0,00	0,00

Fig. 8.3 Example of UTP credit portfolio business plan

Table 8.9 Errors of the model compared to real cases

Portfolio ID	GBV € (~)	Borrower number (~)	Error NPL Vs UTP (%)	Error in strategy det (%)	Error in recovery amount det (%)
P tf1	24,000,000.00	1	0	0	15
ptf2	20,000,000.00	20	10	11	9
ptf3	6,000,000.00	2	0	50	20
ptf4	30,000,000.00	40	6	7	6
ptf5	8,000,000.00	20	4	6	5
ptf6	14,000,000.00	15	9	9	8
ptf7	60,000,000.00	25	11	10	9

References

ATECO, *Update 2022: The Reclassification Mapping—Explanatory Note Source ISTAT* (2007)

L. Bazzocchi, Artificial intelligence and expert systems. Q. Rev. Anal. Critic. (1988)

EBA/GL/2016/07, *EBA Guidelines on the Application of the Definition of Default Pursuant to Article 178 of Regulation (EU) no. 575/2013* (2016).

European Central Bank, *Guidelines for Banks on Non-performing Loans (NPLs)* (2017)

G. Iacono, *The Creation of An Expert System*. francoangeli.it (1991)

ISTAT, *Report on the Competitiveness of the Productive Sectors Ed. 2021*

Market Watch, *Nomisma Elaborations on Banca Ifis Data—Market Watch 2021 and KPMG—Italian Debt Sale Report 2021* (2021)

Osservatorio T6, *Study of the Times of the Courts Ed. 2022*. www.osservatoriot6.com

Provision of the Bank of Italy, *7 June 2017 Statistical and Disclosure Obligations of Securitization Companies, Based on the Provisions of Regulation (EC) No. 1075/2013 of the ECB* (2017)

Regulation (EU) 575/2013, *Capital Requirements "CRR" Regulation, Which Governs the Definition of a Defaulted or "Non-performing" Debtor*

L.K. Woolery, J. Grzymala-Busse, Machine learning for an expert system to predict preterm birth risk. J. Am. Med. Inform. Assoc. **1**(6), 439–446 (1994). https://doi.org/10.1136/jamia.1994.951 53433,PMC116227,PMID7850569

Alessandro Barazzetti Aeronautical Engineer at the Politecnico di Milano, Professor at Ludes Off-Campus Semmelweis University and pilot is the owner of QBT Sagl, a company operating in the development of Algorithms and Software in the Fintech and Proptech sector and of Science Adventure Sagl, a company specialized in the development of applied AI solutions to Health & Care and advanced projects in the Aerospace sector. Alessandro is also founder of SST srl, a company devoted to sustainability studies and certification and Reacorp Sagl a company devoted to Fintech and Proptech services.

Rosanna Pilla is a Lawyer; she has chosen to consolidate her skills in banking and finance law, credit restructuring, executive and repossess procedures. Expert in the active management and enhancement of real estate guarantees, in due diligence and securitisations, credit transfers, preparation and implementation of business plans of secured and unsecured credit portfolios, NPL and UTP credits. Team leader of a group, in constant and dynamic growth, of young professionals who already have various due diligence activities in the in the sale of NPLS credit portfolios, and the establishment of FIA funds. She follows with interest the development and application of new technologies for data analysis and enhancement. Rosanna is a Member of the TSEI Association—Study Table of Italian Executions and founder of Reacorp Sagl, a Swiss company devoted to Fintech and Proptech Services

Chapter 9
More than Data Science: FuturICT 2.0

Federico Cecconi⑩

Abstract In an Asimov novel he imagines that 'machines' can receive input from all the 'facts' that happen in the world, day after day, year after year. Machines are robots in the *Asimovian* sense, and therefore they cannot let their action (or inaction) harm human beings in any way, furthermore they must obey humans and only ultimately protect themselves. And so day after day, year after year, they suggest lines of action, modify procedures, ask for the creation of structures and modify others. In order to promote the welfare of the human race. Without revealing how the story ends (it ends in a very fascinating way anyway), **FuturICT 2.0, a 2019 EU initiative** proposes exactly this, namely the creation of decision-making aid tools, which receive data from the world and propose explanations for dynamics, lines of action, solutions. In this Chap. 1 describe what FuturICT is now, and how this dream hides within it a challenge that is difficult to fully grasp.

Keywords Forecast models · Big data · Social cognition

9.1 What is Data Science?

Starting from the beginning, data science aims to extract knowledge from a huge amount of data by employing mathematics and statistics. And many different techniques are employed to achieve this as good research leads to better results which lead to a profit for the financial institutions. Data science has become extremely relevant in the financial sector, which is mainly used for risk management and risk analysis. Companies also evaluate data models using business intelligence software.

Through the use of Data Science, the accuracy in identifying irregularities and frauds has increased. This has helped reduce risk and scams, mitigate losses and preserve the image of the financial institution. Data science and Finance go hand in hand as Finance is the hub of data. Financial institutions were early pioneers and adopters of data analytics.

F. Cecconi (✉)
LABSS-ISTC-CNR, Via Palestro 32, 00185 Rome, Italy
e-mail: federico.cecconi@istc.cnr.it

© The Author(s), under exclusive license to Springer Nature Switzerland AG 2023
F. Cecconi (ed.), *AI in the Financial Markets*, Computational Social Sciences,
https://doi.org/10.1007/978-3-031-26518-1_9

In other words, data science is an interdisciplinary field that uses scientific methods, processes, algorithms, and systems to extract knowledge rod insights from a lot of structural and unstructured data. Data science is related to data mining, machine learning and big data. Simply put, data science is the collection of data obtained from structured and unstructured sources so as to extract valuable insights. Data sources may include online or manual surveys, retail customer data, and social media usage information and actions. This data is used to model the actions of a network or customer base to predict future behavior and patterns. This research can be extended to a specific sample community, such as retail customers or consumers of social media, weather forecasting, machine learning, and a wide variety of other disciplines.

9.2 What Is the Role of Data Science, for Example, in Finance?

Then, machine learning, big data and artificial intelligence are fascinating and futuristic options for many developers, business people and corporate employees. However, organizations in the financial sector, due to their security concerns, often have a resistance to new technologies. Indeed, the financial world is more driven by cutting-edge developments. While machine learning can streamline loan processes by reducing fraud, AI-powered applications can provide users with advanced recommendations.

Since its inception, data science has helped the evolution of many industries. It is actually one thing that financial analysts have relied on data to derive a lot of useful insights. The growth of data science and machine learning, however, has led to a huge improvement in the industry. Today, automated algorithms and advanced analytical methods are being used together to outpace the competition more than ever before.

To keep up with the latest trends and understand their usage, we will discuss the value of data science in finance by giving many examples. So, data science is used extensively in areas such as risk analysis, client management, fraud detection and algorithmic trading. We will explore each of these and bring you applications of data science in the financial industry.

Initially, data was processed in batches and not in real-time, creating problems for several industries that needed real-time data to gain perspective on current conditions. However, with improvements in technology and the growth of dynamic data pipelines, data can now be accessed with minimal latency.

With this data science in financial application, organizations are able to monitor purchases, credit scores and other financial metrics without latency issues. Financial firms can make assumptions about how each customer is likely to behave based on past behavioral patterns. Using socio-economic apps, they can sort customers into groups and make predictions about how much money each customer expects to receive in the future.

9.3   More than Data Science: FutureICT 2.0

With our knowledge of the universe, we have sent men to the moon. We know micro-scopic details of objects around us and within us. And yet we know relatively little about how our society works and how it reacts to changes brought upon it. Humankind is now facing serious crises for which we must develop new ways to tackle the global challenges of humanity in the 21st century. With connectivity between people rapidly increasing, we are now able to exploit information and communication technologies to achieve major breakthroughs that go beyond the step-wise improvements in other areas.

It is thus timely to create an ICT (Information and Communication Technology) Flagship to explore social life on Earth, and everything it relates to, in the same way that we have spent the last century or more understanding our physical world.[1]

This proposal sketches out visionary scientific endeavours, forming an ambitious concept that allows us to answer a whole range of challenging questions. Integrating the European engineering, natural, and social science communities, this proposal will release a huge potential.

The need of a socio-economic knowledge collider was first pointed out in the OECD Global Science Forum on Applications of Complexity Science for Public Policy in Erice from October 5 to 7, 2008. Since then, many scientists have called for a large-scale ICT-based research initiative on techno-social-economic-environmental issues, sometimes phrased as a Manhattan-, Apollo-, or CERN-like project to study the way our living planet works in a social dimension. Due to the connotations, we use the term knowledge accelerator, here.

An organizational concept for the establishment of a knowledge accelerator is currently being sketched within the EU Support Action VISIONEER, see www.visioneer.ethz.ch. The EU Flagship initiative is exactly the instrument to materialize this concept and thereby tackle the global challenges for mankind in the 21st century.

From the point of view of FuturICT initiative, we have a better understanding of our universe than about the way our society and economy evolve. We have developed technologies that allow ubiquitous communication and instant information access, but we do not understand how this changes our society. Challenges like the financial crisis, the Arab spring revolutions, global flu pandemics, terrorist networks, and cybercrime are all manifestations of our ever more highly interconnected world. They also demonstrate the gaps in our present understanding of techno-socio-economic-environmental systems.

In fact, the pace of global and technological change, in particular in the area of Information and Communication Technologies (ICTs), currently outstrips our capacity to handle them. To understand and manage the dynamics of such systems, we need a new kind of science and novel socially interactive ICTs, fostering transparency, trust, sustainability, resilience, respect for individual rights, and opportunities for participation in political and economic processes.

---

[1] You can see all the information (members, partners, papers, software, solutions) about FuturICT initiative at https://futurict2.eu/.

9.4   An Example: FIN4

One of the reasons for unsustainable behavior is the long period of time between our actions and the consequences they may have. For instance, people have been burning coal for two centuries and driving motor vehicles for one century, while carbon dioxide emissions have only recently been considered to be a serious global problem. Such circumstances create cognitive dissonance, i.e., an uncomfortable feeling of inconsistency. Some researchers propose an innovative way to tackle this problem (see https://link.springer.com/chapter/10.1007/978-3-030-71400-0_4 for details): the creation of FIN4, a socio-ecological finance system.

FIN4 aims to create a self-organizing, highly nuanced incentive system that supports the daily decision-making of people by encouraging desirable actions and discouraging undesirable behaviors.

To promote sustainability, it is important that the FIN4 system takes externalities into account. However, the FIN4 system does this by multiple currencies, reflecting different societal values. By giving externalities prices and allowing them to be traded, they can be internalized to reflect costs and benefits more accurately in a multi-dimensional way. Externalities can be positive (like certain kinds of innovation or infrastructure) or negative (such as pollution).

FIN4 focuses primarily on positive action, rather than on sanctions, for the following reasons:

As an opt-in system, it would be almost impossible to motivate participants to join if they would face a negative balance of tokens, taking into account negative externalities. Obtaining tokens for positive action is a much better value proposition. FIN4, therefore, focuses on rewards for positive actions, rather than on punishments for negative actions.

Our daily choices often seem to be detached from our values. Using a smart distributed incentive system, however, communities can reward what they value—by issuing new types of tokens to address local needs.

These positive actions can be whatever the community considers to be valuable. Examples may include planting trees, using a bike rather than a car, helping a person in need or recycling.

In this case, the tokens can take on any form of value for the communities. They can be fully symbolic (imagine a virtual trophy representing a tree you have planted), give access to a service (such as free bicycle maintenance) or even have a monetary value.

Now, try to understand this by examples. Alice creates a type of token for incentivizing people to pick up and dispose of litter they find in public parks. The community collectively likes this idea and adopts it. While hiking, Bob finds an empty plastic bottle and takes a couple of photos showing where he found it and where he disposed of it. Bob uploads these pictures to the claiming platform and other users approve his claim, rewarding him with a number of positive action tokens.

Sensor or mobile proofs require security measures in the hardware or software to prevent cheating, such as double claiming. Social proofs need an additional incentive layer that motivates other users to provide attestations in good faith, while also preventing collusion.

Placing real-life data on the blockchain is a non-trivial problem and can be susceptible to manipulation or mistakes. In practice, it will likely require combinations of these proofs (like in our example above) in order to adequately prevent cheating. After all, data may not necessarily represent the truth of real-world actions. An oracle based on third-party data might be required to prove that submitted data is correct, before it is irrevocably placed on an immutable blockchain. Once a claim has been proven using these approaches—or indeed combinations of them—the tokens are minted and transferred to the claiming user.

The prover may be the token's creator, a dedicated group of people (such as token holders) or any random user, depending on the nature of the action and type of token. The most effective and useful token systems will likely be approved and adopted by the community in a self-organized and self-sustaining manner.

We envision a multi-layered, multi-dimensional system of decentralized digital cryptocurrencies created at different levels with different characteristics, serving different purposes.

The token systems may operate at a supranational, regional or local level. The different purposes may address environmental, social or other values relevant for sustainability or society.

Any system that allows users to propose new token designs will have to deal with the problem of spam. So, how can we ensure our ecosystem promotes useful token concepts? Rather than establishing rules and barriers to restrict the creation of tokens, our design leverages the innovative capacity of independent token proposals. Every token idea is welcome, but acceptance as an official FIN4 token requires users to vote for it with their governance (GOV) tokens. Thus, all users co-maintain a list of approved positive action tokens.

The reputation tokens (REP) to facilitate social proofs do not yet exist in our system. Their purpose is to help users establish trust in one another in order to interact effectively on the platform. Reputation should reflect the support of the system by the user (e.g., proving, voting, etc.) and not their actual positive action token holdings, which could otherwise introduce bias to the reputation system.

In our current design, users obtain reputation tokens by performing actions that support and strengthen the entire FIN4 system. As a minimum, the actions suitable for obtaining these reputation tokens include (1) the active gathering of tokens (low reward), (2) participation in proof mechanisms (medium reward) and (3) the acceptance of token designs by curators (high reward). Finally, users should also be able to lose these reputation tokens. Also, how can we incentivize entire communities that may already have their own tokens established to join the larger ecosystem? The idea is to represent the additional liquidity won when joining the larger FIN4 network through a "reserve currency" we call liquidity token (LIQ). This token would stand for the network effects gained when enabling larger networks and markets.

One conceivable, yet too simplistic approach would be to create the same number of FIN4 tokens for each accepted token proposal. The overall number of liquidity tokens would, therefore, be commensurate with the total number of FIN4 tokens in the system, thereby giving users more trust in using the different tokens and some assurance that tokens can be exchanged with one another. This could be based on the original idea of "Bancor" by John Maynard Keynes; however, the final design has not yet been decided. Due to the nature of blockchain technology, blockchain creators are unable to prevent trades beyond the confines of the system. Our approach is to create strong incentives to use the platform, while preserving the freedom of users to leave at any time and take their token balances with them. Furthermore, a form of identity is needed when users wish to participate in the governance of FIN4. Identity (ID) here corresponds less to the idea of a scanned passport and more to a concept of self-sovereignty built entirely within the FIN4 system or transferred from other platforms. For example, reputation mechanims may establish identity over time.

9.5   Tackle Complexity[2]

Societies in the twenty-first century have achieved a level of complexity that makes understanding them extremely difficult. Furthermore, any provisional understanding we obtain is frequently surpassed by quick change and new innovations, making social and political responses challenging. Information and communication technology (ICT), which has created new tools for social exchange, has been a major driver of change (e.g., social media, social network sites). Political decision-makers, on the other hand, frequently struggle to keep up with new media developments and to understand their dynamics and impact on society and its many institutions, such as the family, schools and universities, political parties, banks, business organizations, or financial markets. This stems not only from the reality that uneducated and incorrect public policy responses to challenges such as terrorism, food safety, infectious disease outbreaks, financial crises, or environmental disasters kill many persons and incur tremendous financial expenses. It is also necessary to maintain voters' trust and confidence in governments, as poor policies can swiftly erode government legitimacy. World leaders are perplexed by the role of new media in political revolutions like the Arab Spring, social disturbances like the London riots, and terrorist networks. Nonetheless, stakeholders affected by such interventions held decision makers accountable depending on the extent to which they endeavored to make available the best and most comprehensive data and applied the best and most applicable models and theories to make sense of the facts. This is founded on the regulative ideal that we should avoid intervening in a system that we do not fully

---

[2] For a complete description of the issue see Eur. Phys. J. Special Topics 214, 153–181 (2012), FuturICT—The road towards ethical ICT.

understand and cannot foresee how it will respond to our interventions. As long as this regulative ideal continues, the current situation necessitates a greater understanding of society.

The reality that many crises and disasters in modern societies are characterized by a lack of trustworthy real-time data and actionable understanding underscores the need for enhanced understanding. Examples include the 2010 Loveparade disaster in Duisburg, Germany1, an EHEC2 outbreak in numerous European nations in spring 2011, a terrorist attack in Norway in July 2011, and riots in the United Kingdom in August 2011. Such events frequently raise the question, "Could we have known better?" And, more importantly, should we have known better? These are appropriate inquiries, and regrettably, they may have a favorable answer. For example, the work of Helbing and colleagues (and other FuturICT researchers) demonstrates that we can enhance the design of similar events by gaining a greater knowledge of crowd catastrophes.

The ethical conclusion is that a responsible organization and design of future Loveparades necessitates careful consideration of the lessons learnt as well as the science of the relevant phenomena. This example also demonstrates that a crucial role is not only the development of data and scientific understanding, but also the dissemination of this knowledge to decision-makers and players in the actual world. Because ICT has become a component of many social phenomena, there is no hope of understanding the modern information society and human behavior within it without the technology and sciences essential to the study of computer systems and networks.

Then, this is where the fundamental concept of FuturICT comes into play

1. Epistemic Responsibility: Those who are in charge of policies and actions in complex systems are responsible for developing knowledge. The quality of their knowledge tools (i.e., information systems, programs, and data) often determines the quality of their decisions and judgements. Decision-makers' responsibility is thus crucial in terms of designing ex ante epistemic resources and information infrastructures, which is a fundamental goal of FuturICT.
2. Social Knowledge as a Public Good: A wide range of information about society should be available to all citizens under equal opportunity conditions. FuturICT acts as a check on the private sector's development of information monopolies in critical societal domains, contributing to a just and equitable information society.
3. Privacy by Design: In information societies, privacy is a necessary moral restriction for gaining knowledge and comprehension of social reality. Although the phrase refers to a wide variety of moral rights, wants, claims, interests, and duties surrounding (knowledge about) the person, personal lives, and personal identity, privacy is necessary for individual human beings to flourish. Data security technologies must be developed alongside data mining techniques and Esocial science. A key goal of FuturICT is the development of new types of Privacy by Design.
4. Maintaining Trust in the Information Society: Trust involves a moral relationship between the truster and the trustee, a relationship that is partially formed by a belief or expectation that the trustee will act morally. In complex ICT-shaped

environments, trust requires individuals in charge of the environment's design to be as explicit and transparent as possible about the values, principles, and policies that have inspired their creation. This is a fourth guiding concept for FuturICT, whose ultimate goal is a fair information society in which the values, beliefs, and policies that define it are open and transparent.

Federico Cecconi is R&D manager for QBT Sagl (https://www.qbt.ch), responsible for the LABSS (CNR) of the computer network management and computational resources for simulation, and consultant for Arcipelago Software Srl. He develops computational and mathematical models for the Labss's issues (social dynamics, reputation, normative dynamics). For Labss performs both the dissemination that training. His current research interest is twofold: on the one hand, the study of socio-economic phenomena, typically using computational models and databases. The second is the development of AI model for fintech and proptech.

Chapter 10
Opinion Dynamics

Luca Marconi⓪

Managers are not confronted with problems that are independent
of each other, but with dynamic situations that consist of
complex systems of changing problems that interact with each
other. I call such situations messes. Managers do not solve
problems, they manage messes.
Russell L. Ackoff.

Abstract The computational social science is a challenging interdisciplinary field
of study, aimed at studying social phenomena by merging the traditional research
of social science with the use of a computational and data-driven approach. In the
conceptual framework of the complex systems science the environment where these
phenomena live and act can be seen as a social system, composed of many elements,
the agents, which form a social complex network: thus, the agents can be represented by the nodes of a network, whose reciprocal positions determine the resulting
network structure. Therefore, the complex systems science perspective proves particularly useful in order to study the interactions occurring in a social environment: as
a matter of fact, these mechanisms give rise to emergent collective behaviors at
the system level. Hence, the aim of the computational social science is to study
the different interaction mechanisms so as to understand their role in the rise of
the collective social phenomena. In order to perform effective and robust studies,
plenty of models have been proposed, ranging from general social dynamics and
social spreading to crowd behavior and opinion, cultural or language dynamics, or
hierarchy or segregation formation. In this context, opinion dynamics is definitively
a fundamental social behavior to study. By everyday experience, one of the most
important social phenomenon is agreement: especially when it comes to a discussion involving different opinions, the process of reaching at least a partial agreement
proved to be a basic catalyst of the evolution of the human thinking during time.
From the dynamics of the single individuals' interactions to the opinion dynamics in
large populations, there are many situations in the social and political life in which a

L. Marconi (✉)
Advanced School in Artificial Intelligence, Brain, Mind and Society, Institute of Cognitive
Sciences and Technologies, Italian National Research Council, Via San Martino Della Battaglia
44, Rome, Italy
e-mail: luca.marc@hotmail.it

© The Author(s), under exclusive license to Springer Nature Switzerland AG 2023   117
F. Cecconi (ed.), *AI in the Financial Markets*, Computational Social Sciences,
https://doi.org/10.1007/978-3-031-26518-1_10

community needs to reach a consensus. Through a process of progressive agreements on different themes, some opinions spread, some others disappear and some others evolve through merging and non-linear memetic processes: in this way, the dominant macro-culture is formed and evolves in a society. Therefore, opinion dynamics models can be seen as a sub-set of the so-called consensus models. In general, a consensus model allows to understand if and how a set of interacting agents can reach a consensus when choosing among several options: political vote, opinions, cultural features are well-known examples of them. The agent-based modelling methodology proves particularly useful to represent the individuals' behavior and analyze the population dynamics. In the literature, plenty of models are aimed at studying opinion dynamics: each model proposed can often be modified with various variants. In this chapter, we will present the overall scenario of opinion dynamics, by showing the main challenges and approaches to study and analyze consensus formation in heterogeneous settings and conditions.

Keywords Opinion dynamic · Computational social science · Sociophysics

10.1   Introduction to Opinion Dynamics

Opinion dynamics aims at studying, modelling and evaluating in a formal and mathematical the process of the evolution of *opinions* in the course of heterogeneous social *interactions* among *agents* (Xia et al. 2011; Zha et al. 2020). While this field of study is definitively not recent in the literature, its popularity has nowadays grown up: indeed, the increasing interest in *complex systems* and *computational social science* (Lazer et al. 2009; Conte et al. 2012) is definitively stimulating the advancement of such discipline, as well as its application in a wide range of concrete real-world scenarios, e.g. in politics, sociology and also in finance. In this perspective, opinion dynamics strives to model the complex phenomena occurring during the interactions among agents, which give rise *to emerging behaviors* at the system's level: generally, the agents are interconnected by the means of some simple or complex *social network*, allowing some specific interactions, *coordination* and *information exchange* processes (Patterson and Bamieh 2010; Fotakis et al. 2018). Such mechanisms are inherently bounded and limited in their nature, orienting the underlying dynamics and thus allowing scholars and researchers to focus their study towards specific phenomena and situations: in the computational social science perspective, then, the challenge is to effectively model the interaction mechanisms, so as to understand their role in the rise of *collective social phenomena*.

In this conceptual framework, opinion dynamics studies the opinion evolution processes among groups of agents, socially and physically organized in some sort of social network. By everyday experience, one of the most important social phenomenon is the *agreement*, or *consensus*: especially when it comes to a discussion involving different opinions, the process of reaching at least a partial agreement proved to be a basic catalyst of the evolution of the human thinking during time.

Its importance was already pointed out by philosophy and logic, from Heraclitus to Hegel, with the *thesis-antithesis* conceptualizations. There are many situations in the social and political life in which a community needs to reach a consensus. Through a process of progressive agreements on different themes, some opinions spread, some others disappear and some others evolve through merging and non-linear *memetic processes* (Mueller and Tan 2018; Nye 2011): in this way, the dominant macro-culture is formed and evolves in a society. Therefore, the mathematical and statistical study of the different social and cultural phenomena involved in the evolution of opinions in social communities proves to be significant for helping practitioners, decision-makers, managers and stakeholders in heterogenous organizations to improve and optimize decision-making processes under a wide range of potential real-case situations.

In opinion dynamics, the process of designing, adapting, choosing and evaluating the right model under specific static or evolving conditions and boundaries requires a twofold view: on the one hand, it is necessary to model the single, precise, *local* interactions among agents in social groups, usually physically or socially *neighbors* or "similar" to some extent. The concepts of *neighborhood, similarity*, capability or specific *thresholds* to be respected for allowing the interaction should be accurately established and pre-defined. In general, the local interactions among agents are modeled by the so-called *update rules*, encapsulating the specific mechanisms and relationships while taking into consideration the network features and structures where the agents act and interact. Simple, circumscribed local interactions and knowledge exchange processes give rise to *global*, general, emerging behaviors at the whole social network level (Bolzern et al. 2020; Giardini et al. 2015). Thus, one of the main aim of opinion dynamics is to study and analyze the way and the overall time required by a social system to reach the consensus, under several *boundary conditions*. Generally, the final state rends to assume one of three classical *stable states*: consensus, *polarization* and *fragmentation* (Zha et al. 2020). As everybody's daily experience suggests, in social communities or interactions, there is no guarantee that a consensus is always reached: in some cases, two final opinions emerge, polarizing the population to two clusters, or, in other cases, no leading opinions appear, thus the population opinions are fragmented in a certain number of clusters. Therefore, opinion dynamics takes into consideration some specific conceptual elements for constructing effective and complete models: the opinions expression formats, i.e. how the opinions are expressed and modeled (e.g. continuous, discrete, binary etc.), the opinion evolution rules, namely the update rules governing the system dynamics, and the opinion dynamics environment, that is the social network enclosing the agents under specific, circumscribed social or physical relationships.

In this chapter, we will present the overall scenario of opinion dynamics, by showing the main challenges and approaches to study and analyze consensus formation in heterogeneous settings and conditions. We will start by contextualizing the general conceptual framework of computational social science, as a means to effectively study and analyze social phenomena, then we will present a tentative classification of the main opinion dynamics models in the literature, which is useful for our scope. Thereafter, such models will be generally described, with the aim to highlight the main characteristics and behaviors considered. Finally, their advantages and potential applications to the financial domain will be highlighted, in order to help the reader appreciate and understand this fascinating discipline's potentialities for real, concrete and current financial situations and environments.

10.2 The Computational Social Science Perspective

The *computational social science* (CSS) is a challenging and promising highly-interdisciplinary field of study, aimed at studying potentially any *social phenomena* by merging the traditional research of social science with the use of a computational and *data-driven* approach (Lazer et al. 2009; Conte et al. 2012). Then, CSS is focused on allowing social scientists to make use of quantitative, objective, *data-centered* methodologies to integrate their classical and traditional qualitative research with the incoming evidences deriving from quantitative analysis of social phenomena in real-world social contexts and communities. The data exploited to design and evaluate the computational models can come from heterogeneous *data sources*, e.g. social media, the Internet, other digital media, or digitalized archives. Then, CSS is inherently interdisciplinary and multi-faceted: it allows to make inferences regarding human and social behaviors by the application of a wide and evolving range of methodologies and approaches, like statistical methods, *data mining* and *text analysis* techniques, as well as *multi-agent systems* and *agent-based models*. On the other hand, the possible applicative fields range from cultural and political sociology to social psychology, up to economics and demographics.

In the conceptual framework of complex systems science, the environment where these phenomena live and act can be seen as a *social complex system*, composed of many elements, the *agents*, which form a *social complex network*: thus, the agents can be represented by the nodes of a network, whose reciprocal positions determine the resulting network structure. The creation and adaptation of the "right" and most "useful" network structure proves to be a fundamental step in the design and development of CSS methods and algorithms. The different relationships occurring among the nodes can allow to produce some general and emerging behavior in the whole network. Then, the complex systems science perspective proves particularly useful in order to study the interactions generated and evolving in a social environment: these mechanisms give rise to *emergent collective behaviors* at the system level. Thus, some property of the whole network stems from the various relationships occurring among the nodes. Being a complex system, the emergent behaviors cannot be inferred from the single agent behaviors or properties, according to the Aristotle's quote "The whole is more than the sum of the parts" (Grieves and Vickers 2017; Aristotle and Armstrong 1933). In any social system, the interactions among the agents have a pivotal role in all the social processes, ranging from *decision-making* and *collaboration* in organizations, to opinion dynamics and *perception* in social populations, up to *cultural evolution* and *diffusion* in large human societies. Therefore, the main challenge for CSS is to recognize, analyze and predict the evolution of heterogenous social behaviors and phenomena deriving from the social interactions among the agents involved.

Overall, CSS has to deal with two main macro-issues: the *socio-computational* and the *algorithmic-modelling* issues. The former is related to the general needs and challenges related to CSS, e.g. the integration of qualitative social research with data analysis and computational approaches, as well as the need to assess and model the agents' *coordination*, interactions and social processes in different conditions and social situations. The interactions and coordination processes can involve *information* or *knowledge production* and *sharing* among the agents, exactly like in the opinion dynamics models, resulting in the presence or absence of a *final state* at the network level, e.g. some consensus or *synchrony* in the whole system. The latter issues, instead, are related to the specific problems and algorithmic issues related to the design and modelling of the different interaction mechanisms, with the aim to understand their role in the rise of the collective social phenomena. Similarly, the role of information and knowledge creation and exchange have to be algorithmically investigated, due to their significant role in orienting the social and cultural evolution of human populations. The subsequent steps to be faced is the choice, design, implementation and evaluation of the right model, adapted and even extended in the specific case and social dynamics involved.

In the literature, the models proposed deal with many different kinds of social phenomena and processes, ranging from general *social dynamics* (González et al. 2008; Balcan et al. 2009) and *social spreading* (Conover et al. 2011; Gleeson et al. 2014), to *crowd behavior* (Helbing 2001) and opinion dynamics (Holley and Liggett 1975; Granovsky and Madras 1995; Vazquez et al. 2003; Castellano et al. 2003; Vilone and Castellano 2004; Scheucher and Spohn 1988; Clifford and Sudbury 1973; Fernández-Gracia et al. 2014), *cultural dynamics* (Axelrod 1997), *language dynamics* (Baronchelli et al. 2012; Castelló et al. 2006), *hierarchy* (Bonabeau et al. 1995) and *segregation formation* in social populations. The general macro-approach involved exploits the dichotomy between *micro* and *macro* levels in the complex social system considered: the models generally focus on few and simple mechanisms included in the system, orienting the local interactions among the agents, so as to study the variation of the social dynamics in response to basic interaction rules, thus eliciting the single contribution of a specific phenomenon to the collective behavior of the social system. By the specific inclusion and adaptation of social mechanisms at the local level, the models are able to reflect the increasing complexity of real-world social situations, allowing practitioners and scholars to both focus on some specific mechanism and assess the overall emerging behaviors and phenomena in the global network.

In this general perspective, now we proceed by focusing on opinion dynamics and the main models presented in the literature. We will start by depicting the general scenario of opinion dynamics, by showing a possible classification of the different approaches in the state-of-the-art. Then, we will follow the proposed classification strategy to focus on the different models and mechanisms, to make the reader appreciate their relevance in social settings and, finally, in concrete financial applications.

10.3 Classification of Opinion Dynamics Models

Opinion dynamics models are widely diffused and popular in the literature, in the conceptual and disciplinary framework of computational social science, mathematical and *statistical physics-based* modelling of human behavior (Castellano et al. 2009; Perc et al. 2017), as well as complex systems. The rise, evolution, adaptation and modifications of models are definitively ongoing processes in the state-of-the-art, thus providing a comprehensive and complete classification is challenging. To our scope, we aim to allow reader to orient himself or herself in the set of traditional, basic and fundamental models we will report in the next sections. The final aim is to let the reader appreciate their specific advantages or potential applications when it comes to finance and its challenges. Therefore, this work is surely not intended as an exhaustive review of the models in the literature.

To provide a useful classification for our scopes, we consider two main dimensions: *dimensionality* and *tipology* of opinions. Indeed, on the one hand opinions can be represented by *one-dimensional or multi-dimensional* vectors (Sîrbu et al. 2017); on the other hand, the vectors modelling the opinions can be either *discrete* or *continuous* (Zha et al. 2020; Sîrbu et al. 2017). Multi-dimensional opinions can be useful when modelling agents characterized by a set of *characteristics* or *features* of their whole opinions. This happen in the case of the Axelrod model (Sîrbu et al. 2017), where the agents are endowed with different cultural features modeling the different believes or attitudes of the individuals in the population. Instead, one-dimensional, classical opinion dynamics model are widely used to study opinion dynamics in different conditions and population, focusing on specific kinds of opinions and on their evolution according to the rules governing the system. Moreover, opinions can be either discrete or continuous. In the former case the components of the opinion vector can only assume discrete values, thus encapsulating a finite number of possible states, while in the latter case the components of the opinion vectors are continuous, assuming values in the domain of real numbers (with any specific constraints, if needed). Finally, in the literature there are also hybrid models, where the different possibilities are mixed according to the specific social phenomena or behaviors to study and simulate, e.g. in the CODA (continuous opinions and discrete actions) model (Zha et al. 2020; Sîrbu et al. 2017) we will report.

Of course, there could be several other ways to provide a classification methodology for studying and appreciating the different opinion dynamics models in the literature. In particular, it could be possible to consider other significant differentiation elements, specifically the types of *stable states* reached in the consensus formation process, as well as the agents' behaviors, the social phenomena studied, or even the kind of social network exploited. In any case, it could be possible to construct arbitrary classifications or categorizations of the many models in the literature by focusing on the presence or absence of specific modelling factors. Notably, the role of *information representation* and *information sharing* could be fundamental to further explore the possible classifications of opinion dynamics models, especially when it comes to the role of *external information*, namely deriving from information

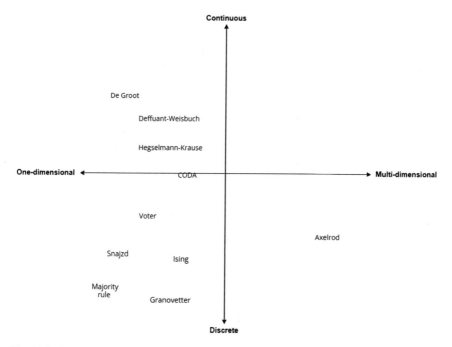

Fig. 10.1 Classification positioning map of the main opinion dynamics models

sources that are external respect to the system population (e.g. mass media broadcast). Nevertheless, we decided to make the classification chosen as simple as possible, to focus on the main fundamental models and their application in finance, as anticipated before.

In Fig. 10.1, we schematically represent the different models according to the presented classification.

10.4   Continuous Opinion Models

Let us start by presenting the main opinion dynamics models with one-dimensional and continuous opinions. These models are widely exploited in the literature to analyze and stimulate real-world scenarios, where opinions related to a specific subject (in a one-dimensional way) are not discrete or binary, but they vary on a continuous scale, with specific boundaries or limitations when needed. In the literature, the main models in such category are the De Groot model (Degroot 1974; Das et al. 2014) and the bounded confidence models: the Deffuant-Weisbuch model (Weisbuch et al. 2002; Deffuant et al. 2000) and the Hegselmann-Krause model (Hegselmann and Krause 2002).

De Groot Model

This is a classical opinion dynamics models, widely studied as a basic approach for more advanced models or extensions. According to De Groot, the *opinion formation* process and their dynamics are defined according to a *linear combination* of the initial opinions of the interacting agents. Then, let us consider the set of the agents $A = \{a_1, a_2, \ldots, a_m\}$ and let $o_i^t \in R$ be the opinion of agent a_i at round t. As the model uses a linear combination for defining the evolving opinions of the agents, we introduce the weight w_{ij} the agent a_i assigns to agent a_j, where the conditions $w_{ij} \geq 0$ and $\sum_{j=1}^{m} w_{ij} = 1$ must be satisfied. Then, the De Groot model exploits the following update rule, governing the evolution of the opinion of agent a_i:

$$o_i^{t+1} = w_{i1}o_1^t + w_{i2}o_2^t + \cdots + w_{\mathfrak{z}}o_m^t \, t = 0, 1, 2, \ldots$$

This update rule is definitively equivalent to:

$$O^{t+1} = W \times O^t t = 0, 1, 2, \ldots$$

Or, alternatively:

$$O^t = W \times O^{(t-1)} = W^2 \times O^{(t-2)} = \cdots = W^t \times O^0,$$

where $W = (w_{ij})_{m \times m}$ is a row-stochastic weight matrix encapsulating the different weights employed by the agents, where W^t is its t-th power, and $O^t = (o_1^t, o_2^t, \ldots, o_m^t)^T \in R^m$. This model is inherently linear and it is traceable over time. It is an iterative averaging model and the consensus can be reached: according to De Groot, if there is a time t such that every element in at least one column of the matrix W^t is positive, then it is possible to reach the consensus. In other words, a consensus can be reached if there exists a power t of the weight matrix when W^t has a strictly positive column. Due to its simplicity and linearity, the De Groot model presents several extensions in the literature: these are the so-called DeGrootian models (Degroot 1974; Noorazar et al. 2020; Noorazar 2020). One of the most diffused ones is the *Friedkin-Johnsen* model (Noorazar 2020; Friedkin and Johnsen 1990), which include *stubborn agents* in the system, by considering a parameter representing the *susceptibility* of each individual agent, namely the level of *stubbornness*.

Bounded Confidence Models

The bounded confidence models can be seen as extensions of the De Groot model, where the agents interact according to a given *threshold* of the distance between their opinions and where the weights change over time or with the opinions of the agents. The main idea is the relevance of the *distance* between the opinions as determining the occurrence of the interactions: if the difference between two considered agents is below a given threshold, then the agents will interact and their opinions will possibly change, otherwise they definitively do not consider each other because the two ideas are too far from each other. If the threshold assumes the same value for all the agents, then the bounded confidence model is *homogeneous*; otherwise, the model is *heterogeneous*. The two main bounded confidence models are the Deffuant-Weisbuch model, wich is a *pairwise* model, and the Hegselmann-Krause model, which is the mostly diffused *synchronous* version.

Deffuant-Weisbuch Model

In the Deffuant model, at each discrete time step two neighboring agents are randomly selected from the agent set $A = \{a_1, a_2, \ldots, a_m\}$ and they interact in a pairwise manner. Only if their opinion difference is below a given threshold ε, the result of the interaction is a sort of compromise toward each other's opinion. Otherwise, there is no modification in their opinion. Then, if $\left| o_i^t - o_j^t \right| \leq \varepsilon$, the agents will interact and change their opinions, otherwise they will not interact. In the case the interaction happens, the update rule for the agent i is the following one:

$$o_i^{t+1} = o_i^t + \mu\left(o_j^t - o_i^t\right),$$

while the update rule for the agent j is:

$$o_j^{t+1} = o_j^t + \mu\left(o_i^t - o_j^t\right).$$

Then, in this model there are two main parameters to take into consideration: the *convergence parameter* μ, also called *cautiousness parameter*, which governs the speed of convergence, or, in other words, the capability of an agent to be influenced by the opinion of another agent in the population; and the *confidence radius* ε, namely the threshold governing the actual capability of an agent to interact with another agent in the population. The Deffuant-Weisbuch model is extensively studied in many conditions related to the two parameters. According to the different situations and values assumed by the two parameters, the model can lead to the three possible stable states: consensus, polarization and fragmentation. In particular, in the special case $\mu = \frac{1}{2}$, the outcome of the interaction will be a complete agreement at the average of their previous opinions. Several extensions are found in the literature, considering aspects like the presence of *noise*, the kind of agents interacting and specific boundaries or conditions on the values of the parameters.

Hegselmann-Krause Model

This is the most well-known and diffused synchronous version of bounded confidence model. Again, there are many versions of this model. Our aim is just to present the most simple and used one, so as to allow the reader to appreciate the potentialities of this approach for real-world applications. The most simple update rule here is given by:

$$o_i^{t+1} = \frac{1}{\left|N_i^t\right|} \sum_{j \in N_i^t} o_j^t$$

where N_i^t is the set of neighbor of agent i at time j, including the agent i itself, where the condition $\left| o_j^t - o_i^t \right| \le r$ is respected, namely where, again, the difference between the opinions of the possibly interacting agents is below a given threshold. This model is studied with specific attention to the neighbors' opinion dynamics and their convergence properties, even in an extended multi-dimensional case.

10.5 Discrete Opinion Models

These models are widely used to explore and analyze real-life situations where opinions assume a discrete range of possible states. In general, many models assume binary opinions, at least in their initial version of the model, with several extensions to multiple discrete opinions in further versions. Often such models are exploited to study and simulate political scenarios, as in the case of Trump's victory. Again, our aim here is not to cover all the models designed and applied in the literature of opinion dynamics, rather to present the most important and basic models used as a basis for more complicated and advanced approaches: the idea is to introduce the reader to the many potentialities of these approaches for finance, which we will explore in the last section of this chapter.

Voter Model

The voter model is one of the most classical and simplest discrete opinion dynamics models. As the name suggests, it has been widely applied to the study of electoral competitions and political systems. Let us consider a population of c agents and let $A = \{a_1, a_2, \ldots, a_n\}$ be the set of the agents in the population, with o_i^t is the opinion of the agent i at time t, assuming a binary state, namely $o_i^t = 1$ or $o_i^t = -1$. Then, at each time t a random agent a_i is selected. Then, a second random agent a_j is chosen and the opinion o_i^{t+1} of agent a_i will assume the value of o_j^t, namely $o_i^{t+1} = o_j^t$. In this way, after the interaction the agent i effectively adopts the state of the randomly chosen neighbor. In this model, there are two possible final consensus states and the final state reached in the population depends on the initial distributions of the opinions of the agents. This model has been widely and extensively studied in populations of agents located in two-dimensional regular lattices. Moreover, many studies are related to d-dimensional hyper-cubic lattices. In an infinite system, it has been proved that consensus is reached for $d \le 2$.

Sznajd Model

The idea behind the Snajzd model can be effectively conveyed by the simple motto "united we stand, divided we fall". In other words, this model encapsulates two specific social behaviors that can occur in social systems, at the neighborhood level: *social validation* and *discord destroys*. The former regards the propagation of a specific opinion among neighbors: if two (or more) agents in the neighborhood share the same opinion, then their neighbors will agree with them, i.e. the opinion will propagate and become dominant in this neighborhood. The former, instead, regards the fragmentation of the opinions in the neighborhood: if two (or more) agents disagree, namely they have opposite opinions, then their neighbors will argue with them, as each agent will change their opinion until they will not agree with their adjacent neighbor. In a more formal way, let $o_i^t = 1$ or $o_i^t = -1$ be the opinion of the agent a_i at time t. The update rules governing the evolution of the opinions are the following:

1. At each time t a pair of agents a_i and a_{i+1} are randomly chosen and their neighbors are a_{i-1} and a_{i+2}.
2. If a_i and a_{i+1} share the same opinions, it will be propagated to a_{i-1} and a_{i+2}, i.e. if $o_i^t = o_{i+1}^t$ then $o_{i-1}^{t+1} = o_{i+2}^{t+1} = o_i^t$.
3. If a_i and a_{i+1} do not share the same opinions, the different opinions will be propagated to the agent neighbors, namely if $o_i^t = -o_{i+1}^t$ then $o_{i-1}^{t+1} = o_{i+1}^t$ and $o_{i+2}^{t+1} = o_i^t$.

Ising Model

The Ising model is a well-known and extensively studied model in physics, one of the most classical approaches in statistical mechanics to analyze the *phase transitions* in ferromagnetic materials. In the field of opinion dynamics, it has been studied to model social interactions in situations where the agents have binary opinions, represented by *spins*, which can be *up* or *down*. This choice can seem inherently limited, but it proved to be useful in real-life situations where people have to deal with a choice between two opposite options regarding a specific topic. Spin couplings are used to represent peer interactions among the agents. Furthermore, the model also considers another element, which is external information (e.g. media promotion), represented by the magnetic field in the model. Thus, the energy of interaction, called $E = -Jo_io_j$ represents the degree of conflict between the agents, where $o_i = \pm 1$ and $o_j = \pm 1$ represent the agents' spins, corresponding to states up and down, respectively, and J is an energy coupling constant, denoting the interaction strength among agents. Moreover, there is an interaction energy $E = -Ho_i$ between the opinions and the external information. The former interaction energy will be minimal when two agents share the same opinions, while the latter is minimized when the opinion and the external field share the same value.

Majority Rule Model

This model was introduced by Galam to study the use of majority rule in the evolution of opinion dynamics in social systems. In this case, a randomly selected group of r agents is chosen in the agent population, with all the opinions being binary. Then the majority rule is employed to update the opinions and, finally, all the process starts again with other randomly chosen agents. In the literature, many versions of the update rules and conditions has been studied. In the Galam version, the opinion o_k^{t+1} of the agent a_k at the time $t + 1$ will be equal to 1 if $\sum_{i=1}^{r} o_i^t/r > 0.5$, while $o_k^{t+1} = o_k^t$ if $\sum_{i=1}^{r} o_i^t/r < 0.5$ and $o_k^{t+1} = 0$ in all the other possible cases. The work by Galam was significant to enhance and stimulate the study of opinion dynamics in hierarchical systems and decision-making in political and electoral systems in several conditions, including the presence of coalition or fragmentation between opinions among agents in the population.

Granovetter's Threshold Model

This is one of the most famous threshold models, investigating the possible presence and role of a critical threshold in the number of agreeing decision-makers, which, if reached, determines the realization of a consensus. In this model, each agent has a certain activation threshold determining his or her choice of actively participating within a collective behavior, e.g. a specific political decision or decision-making process. This threshold basically depends on the number of other agents that the specific individual should observe participating before deciding to join the group and the process. Then, a population of N agents, each individual is provided with a particular threshold defining the number of participating agents needed before the considered individual decides to take part to the decision-making process and collective behavior. The study of such thresholds in heterogeneous conditions and populations allows to predict the final state of the opinions in the population.

10.6 Hybrid Models

Besides discrete and continuous classical opinion dynamics models, in the literature it is possible to find several hybrid approaches, mixing different kinds of opinions or merging heterogeneous models, with the aim to mutually reinforce the methods included and exploit their synergies. The hybridization of different model can be either a way to study certain behaviors or mechanisms in a larger and more comprehensive framework, also to effectively create a method able to better simulate them in a real-world application or scenario. In this short section, we briefly present the Continuous Opinions and Discrete Actions (CODA) model, providing a explanatory approach for hybridization of discrete and continuous elements in opinion dynamics.

CODA Model

In this model, the agents deal with binary decisions, thus representing the (very usual) situation where an individual faces two conflicting opposite opinions to choose. Then, each agent i has to pick one choice between two discrete (binary) opinions. In this scenario, it is reasonable to assume that each agent i would choose one of the two alternatives with a probability p_i. The CODA approach helps to model such probability, allowing to model the degree of choosing a certain discrete opinion, under specific conditions. Then, we have a variable representing the two opposite choices, namely $x = \pm 1$, and all the agents are represented on a square lattice with a continuous probability p_i denoting the degree of agreement to the opinion $+1$, while the probability $1 - p_i$ corresponds to the extent of agreement to the opinion -1. The agents choose a discrete opinion by using a threshold $\sigma_i = sign(p_i - 1/2)$. Moreover, each agents states which choice is preferred, thus informing the others. All the agents only see the discrete opinion of the others and change the corresponding probability according to their neighbors, by the means of a specific Bayesian update rule, with the aim to stimulating and favoring agreement in the neighborhood. The interaction among the neighbors can happen according to different schemas, either a Voter model one, where each agent interacts with one neighbor at each step, or a Sznajd model one, where two neighbors are able to influence the other ones. In any cases, the model shows the emergence of extremism even without extreme opinions at the beginning. Several extensions of the model have been studied, e.g. the presence of migration in the agents social network, namely the possibility for the agents to move and change their position respect to the other agents in the network. Other cases include the inclusion of a third possible opinion allowed, as well as the possibility to provide the agents with some "awareness" of the effect they produce on the other individuals. Finally, it is worth to mention the possibility to include some "trust" in the system, by making each agent able to judge the other ones, categorizing them into "trustworthy" and "untrustworthy". In each case, the possible emergence of consensus or polarization is studied under several heterogeneous conditions.

Besides the CODA model, many other hybrid approaches have been investigated and explored in the literature, ranging from models taking into account also the past relationships of the agents during their interaction, to attempts of creating hybrid continuous-discrete dynamics, even to models striving to include cognitive or psychological phenomena or behaviors into the structure or into the interactions among the agents. In order to fully exploit the potentialities of some models, sometimes they are extended to multiple dimensions of opinions. Indeed, until now we only considered one-dimensional models, which prove to be effective in many real-world scenarios and concrete applications. Nevertheless, many situations require more than one dimensions to adequately represent the full range of the opinions considered: thus, it is worth to mention multi-dimensional opinion dynamics models.

10.7 Multi-dimensional Opinions

In the widespread scenario of opinion dynamics models and approaches, it is necessary and worth to mention the multi-dimensional case, enabling the design, development and application of several methods, either specifically created for these situations or extended from the one-dimensional case. At a first glance, such models could appear somehow strange or useless in real-world scenarios. Instead, such models are particularly appliable and necessary when the opinions of the agents in the population are composed and influenced by a series of heterogeneous variables, representing their features or characteristics, related to different sociological, political or cultural traits, influencing and composing the whole opinion. Then, the use of multi-dimensional models can be even more realistic or useful than for real-life complex scenarios, where agents, their opinions and characteristics are complex and related to several dimensions. Consequently, plenty of studies in the literature are focused on exploring such models, often extending and adapting the one-dimensional classical cases, as it happens for bounded confidence models. Among the aspects that are particularly studied in the literature, there are the change in *confidence levels*, the model *convergence, robustness* and *stability* in heterogeneous conditions influencing the appearance of consensus, polarization or segmentations of the opinions in the population. In this short section, we briefly present the main aspects of the well-known Axelrod model, where the agents are endowed with different cultural features modeling their beliefs or attitudes.

Axelrod Model

This model was introduced for culture dynamics, to study and simulate the culture formation and evolution processes. In the population, two main sociological phenomena are included, namely homophily and social influence. The former refers to the preference for interacting with similar peers, thus agents characterized by a set of similar features. Homophily is a common sociological phenomenon and it can lead to the emergence of clusters in the population of the interacting agents, in many kinds of social networks and social communities. The latter, instead, refers to the tendency of the individual to become more similar after their interaction, thus it is the ability to mutually influence each other in the curse of the interaction.

In the population of N individuals, each agent has a culture modeled by F variables, thus there is a vector $(\sigma_1, \ldots, \sigma_F)$ of cultural features. Each of them can assume q discrete values, which are the traits of the different features, where $q_f = 0, 1, \ldots, q - 1$. This structure is definitely useful to effectively and immediately model a set of different beliefs, behaviors or attitudes of the agents. An interaction between two agents i and j occurs according to the probability of interaction $o_{ij} = \frac{1}{F}\sum_{f=1}^{F}\delta_{\sigma_{f(i)},\sigma_{f(j)}},$ where δ is Kronecker's delta. Then, according to such probability, if the interaction happens, one of the features is selected and $\sigma_{f(j)} = \sigma_{f(i)}$ is set. Otherwise, the interaction does not occur and the two agents maintain their features, respectively. This model lead a randomly-initialized population to two possible absorbing states: either a sort of consensus, namely an ordered states where the agents share traits, or a sort of segmentation, where multiple clusters of different traits co-exist in the population. The final state specifically depends on the initial number of the possible q initial traits: if q is small, the interaction among agents with similar traits is easier, thus consensus can be reached. Instead, for larger q, there are not enough similar traits shared, thus the interaction between similar agents is not ale to create and maintain growing cultural domains. Of course, the study of the phase transition between the two possible states is particularly significant. On regular lattices, such phase transition appears at a critical q_C, depending on the number F of the features in the population.

Of course, such model has been extensively analyzed, adapted and extended in the literature. Overall, it has a simple rather significant structure for studying both cultural dynamics and opinion dynamics at the presence of multi-dimensional variables able to describe complex opinions or cultural features. Thus, such model lends itself to many possible adaptations, modifications or even extensions to more general structures. Among the possible modified versions, it is relevant to mention the cases of modification of the update rule, e.g. always allowing to choose a pair of interacting agents, instead of randomly selecting them. Other interesting models exploit the so-called *cultural drift*, namely external *noise* influencing the formation of cultural clusters and the appearance of consensus. Finally, it is also possible to change the attitudes of the agents toward the change in their opinion, e.g. including some threshold for the overlap, determining when the agents will disagree, as well as including *committed individuals* that will not change their opinion in any case.

10.8 Applications to Finance

Opinion dynamics is widely applied in finance, in business and organizational studies. Let us remind that the financial market is inherently a complex systems, where a large set of human agents interact, exchange information, trade and create groups with internal dynamics. This is the basic concept to take into consideration when striving to comprehend the many possible applications of opinion dynamics and, in general, mathematical modeling of social behavior, to the financial domain. Moreover, it is also necessary to recall that such agents are endowed with heterogeneous properties, e.g. related to their *cultural, sociological* or *demographic* features, as well as different *strategies of investments* and *risk propensity*. These characteristics and behaviors at the *micro-* and *meso*-level give rise to emergent phenomena, influencing the evolution of the financial market and, in general, of the whole society. In this context, opinion dynamics proved to be a useful methodological and mathematical tool to model the structure of social networks in the financial domain, simulate the range of possible behaviors and study emergent and collective phenomena in real-world situations. While in the financial situations it is common to exploit binary models, allowing to study the cases where the agents have to choose between two opposite actions, it is generally possible to find applications of the already presented basic models or their extensions, as well as newly created models, specifically designed to analyze particular situations or scenarios.

Among the models presented, the Ising one is one of the most commonly applied to the financial domain, both in its original version and in specifically designed extensions for finance. In this case, it is necessary to limit the study to the binary case, which is not a problem as this is a typical scenario in the real-world applications of opinion dynamics and in finance. The Ising model can be used to model the interactions and *trading strategies* among *bullish* and *bearish* traders, in an environment acting with external forces influencing the interactions and behaviors of the agents. Moving a step forward, there are extensions able to better represent the forces at work in this case: in particular, the Ising model basic elements can be extended to also take into consideration the so-called *mutual influences*, namely the *interaction strength* or the influence of the decision of a trader to some other one, but also the *external news* or other possible external forces influencing the trader decisions and possible trader characteristics or traits. Other possible extensions of the Ising model enlarge the range of the potential trader choice, allowing the agents to decide whether to buy, sell or not to act. Therefore, the Ising model proved to be useful both for modeling the agents features and the forces influencing their interactions, as well as to model the kind of actions the traders can perform, thus giving rise to a wide set of possible simulations of financial situations and scenarios.

The Sznajd model can allow to better understand and study some specific characteristics of the opinion dynamics in the case of trader groups. Indeed, such groups, as many other social systems, are affected by *herding* behaviors, where some *followers* emulate the decisions of a *leader*. As the Sznajd model represents the social phenomena of *social validation* and *discord destroys*, this method can encapsulate some common and typical social mechanism that usually are present in the financial markets. Indeed, when there are some *local gurus* with the authority, the experience and the communication ability and channels to influence its local neighbors, then many market participants in that group are *trend followers*: they will make their decisions and order placements according to the opinion of their local guru. On the contrary, when such gurus are not present in the neighborhood, the trend followers will act randomly and there is not a dominant opinion prevailing in the system. Moreover, there could be *rational traders* added to the model: these traders have an exact knowledge of the level of demand and supply in the market, thus they are able to calculate the way to exactly place their orders.

Similarly, the voter model was applied to the financial domain to study the social behaviors in local neighborhoods when it comes to the diffusion and evolution of financial choices and actions among traders. The possible applications regard the use of the voter model to study the states of the agents on the market: buy, sell or no action. Other applications study the *entropy fluctuations behaviors* in the financial market by the means of voter-based models. One of the most significant applications of the voter model employs the so-called *social temperature*, which represents the market temperature, influencing the *persuasiveness* of some agents groups and, subsequently, the uncertainty of the agents' investment strategies.

The use of the basic models is significant to analyze the behaviors of the agents in complex financial and social environments, thus allowing to simulate and understand the market evolution, as well as phenomena like *market trends*, particular *bubbles* or *crashes* occurring in the market history. Thus, such models are able to let the researchers and scholars appreciate and examine the different kinds of *phase transitions*, *stable states* and the general evolution of the markets under several heterogeneous conditions. Nevertheless, there are many possible extensions of the basic models, as well as specifically designed models to capture some particular financial situations and mechanisms or aspects. One primary extension is the use of continuous models, instead of the simple and common discrete binary opinions and time. In addition, the level of complexity can be increased by considering both opinion and *information exchange* among the agents. As the structure of the models include more elements than in the most classical and basic methods, it can be appropriate to exploit *analytical* solutions of the models, based on the use of *differential equations*, instead of the numerical and computer simulations. All these possibilities are generally present in the so-called *kinetic models*, which are used for studying the evolution of speculative markets, or the mechanism of *price formation* and changing and risk propensity. Instead, other models try to merge the agent-based and classical opinion dynamics methodology to the use of machine learning, in a data-driven and more computational way. Such synergy gives rise to probabilistic models to study opinion dynamics by the means of the use of a wide set of probabilistic parameters, which can be useful to study the opinion evolution processes in complex stochastic situations. From the analytical point of view, in this case often the *stochastic differential equations* are employed, as well as neural network structures, to help study the nonlinear dependencies and opinion dynamics in the social networks.

Overall, opinion dynamics is a highly promising methodological tool to help scholars, researchers and practitioners to deeply understand the market structures and evolution, to capture relevant and hidden social and financial phenomena and predict the future market states in a wide range of potential applications and situations. We are convinced that further studies of such methods, especially in synergy with the many computational potentialities of machine learning and deep learning, will shed light on many social and financial mechanisms, thus allowing to take a step forward toward the comprehension of one of the more fascinating complex systems, deeply affecting our everyday life.

References

H.T. Aristotle, G.C. Armstrong, *The Metaphysics* (W. Heinemann, G. P. Putnam's Sons, London, 1933)

R. Axelrod, The dissemination of culture: a model with local convergence and global polarization. J. Conflict Resolut. **41**(2), 203–226 (1997)

D. Balcan, V. Colizza, B. Gonçalves, H. Hu, J.J. Ramasco, A. Vespignani, Multiscale mobility networks and the spatial spreading of infectious diseases. Proc. Natl. Acad. Sci. **106**(51), 21484–21489 (2009)

A. Baronchelli, V. Loreto, F. Tria, Language dynamics. Advs. Complex Syst. **15**, 1203002 (2012)

P. Bolzern, P. Colaneri, G. De Nicolao, Opinion dynamics in social networks: the effect of centralized interaction tuning on emerging behaviors. IEEE Trans. Comput. Soc. Syst. **7**(2), 362–372 (2020)

E. Bonabeau, G. Theraulaz, J.L. Deneubourg, Phase diagram of a model of self-organizing hierarchies. Phys. A **217**(3), 373–439 (1995)

C. Castellano, D. Vilone, A. Vespignani, Incomplete ordering of the voter model on small-world networks. EPL (Europhys. Lett.) **63**(1), 153 (2003)

C. Castellano, S. Fortunato, V. Loreto, Statistical physics of social dynamics. Rev. Mod. Phys. **81**(2), 591 (2009)

X. Castelló, V.M. Eguíluz, M. San Miguel, Ordering dynamics with two non-excluding options: bilingualism in language competition. New J. Phys. **8**(12), 308 (2006)

P. Clifford, A. Sudbury, A model for spatial conflict. Biometrika **60**(3), 581–588 (1973)

M. Conover, J. Ratkiewicz, M.R. Francisco, B. Gonçalves, F. Menczer, A. Flammini, Political polarization on Twitter. ICWSM **133**, 89–96 (2011)

R. Conte, N. Gilbert, G. Bonelli, C. Cio-Revilla, G. Deuant, J. Kertesz, V. Loreto, S. Moat, J.-P. Nadal, A. Sanchez, A. Nowak, A. Flache, M. San Miguel, D. Helbing, Manifesto of computational social science. Eur. Phys. J. Special Top. **214**, 325346 (2012)

A. Das, S. Gollapudi, K. Munagala, Modeling opinion dynamics in social networks, in *Proceedings of the 7th ACM international conference on Web search and data mining (WSDM'14)* (Association for Computing Machinery, New York, NY, USA, 2014), pp. 403–412. https://doi.org/10.1145/2556195.2559896

G. Deffuant, D. Neau, F. Amblard, G. Weisbuch, Mixing beliefs among interacting agents. Adv. Complex Syst. **3**(01n04), 87–98 (2000)

M.H. Degroot, Reaching a consensus. J. Am. Stat. Assoc. **69**(345), 118–121 (1974). https://doi.org/10.1080/01621459.1974.10480137

J. Fernández-Gracia, K. Suchecki, J.J. Ramasco, M. San Miguel, V.M. Eguíluz, Is the voter model a model for voters? Phys. Rev. Lett. **112**(15), 158701 (2014)

D. Fotakis, V. Kandiros, V. Kontonis, S. Skoulakis, Opinion dynamics with limited information, in *International Conference on Web and Internet Economics* (Springer, Cham, 2018), pp 282–296

N.E. Friedkin, E.C. Johnsen, Social influence and opinions. J. Math. Sociol. **15**(3–4), 193–206 (1990)

F. Giardini, D. Vilone, R. Conte, Consensus emerging from the bottom-up: the role of cognitive variables in opinion dynamics. Front. Phys. **3**, 64 (2015)

J.P. Gleeson, D. Cellai, J.P. Onnela, M.A. Porter, F. Reed-Tsochas, A simple generative model of collective online behavior. Proc. Natl. Acad. Sci. **111**(29), 10411–10415 (2014)

M.C. González, C.A. Hidalgo, A.L. Barabási, Understanding individual human mobility patterns. Nature **453**(7196), 779–782 (2008)

B.L. Granovsky, N. Madras, The noisy voter model. Stoch. Proces. Appl. **55**(1), 23–43 (1995)

M. Grieves, J. Vickers, Digital twin: mitigating unpredictable, undesirable emergent behavior in complex systems, in *Transdisciplinary Perspectives on Complex Systems* (Springer, Cham, 2017), pp. 85–113

R. Hegselmann, U. Krause, Opinion dynamics and bounded confidence models, analysis, and simulation. J. Artif. Soc. Soc. Simul. **5**(3), 1–33 (2002)

D. Helbing, Traffic and related self-driven many-particle systems. Rev. Mod. Phys. **73**(4), 1067 (2001)

R.A. Holley, T.M. Liggett, Ergodic theorems for weakly interacting innite systems and the voter model. Ann. Probab. 643–663 (1975)

D. Lazer, A. Pentland, L. Adamic, S. Aral, A.L. Barabasi, D. Brewer, N. Christakis, N. Contractor, J. Fowler, M. Gutmann, T. Jebara, G. King, M. Macy, D. Roy, M. Van Alstyne, Life in the network: the coming age of computational social science. Science **323**(5915), 721723 (2009)

S.T. Mueller, Y.Y.S. Tan, Cognitive perspectives on opinion dynamics: The role of knowledge in consensus formation, opinion divergence, and group polarization. J. Comput. Soc. Sci. **1**(1), 15–48 (2018)

H. Noorazar, Recent advances in opinion propagation dynamics: a 2020 survey. Eur. Phys. J. plus **135**(6), 1–20 (2020)

H. Noorazar, K.R. Vixie, A. Talebanpour, Y. Hu, From classical to modern opinion dynamics. Int. J. Mod. Phys. C **31**(07), 2050101 (2020)

B.D. Nye, *Modeling Memes: A Memetic View of Affordance Learning.* Doctoral dissertation, University of Pennsylvania (2011)

Patterson S, Bamieh B (2010) Interaction-driven opinion dynamics in online social networks, in *Proceedings of the First Workshop on Social Media Analytics*, pp. 98–105

M. Perc, J.J. Jordan, D.G. Rand, Z. Wang, S. Boccaletti, A. Szolnoki, Statistical physics of human cooperation. Phys. Rep. **687**, 1–51 (2017)

M. Scheucher, H. Spohn, A soluble kinetic model for spinodal decomposition. J. Stat. Phys. **53**(1), 279–294 (1988)

A. Sîrbu, V. Loreto, V.D. Servedio, F. Tria, Opinion dynamics: models, extensions and external effects, in *Participatory Sensing, Opinions and Collective Awareness.* Springer, Cham (2017), pp. 363–401

F. Vazquez, P.L. Krapivsky, S. Redner, Constrained opinion dynamics: freezing and slow evolution. J. Phys. a.: Math. Gen. **36**(3), L61 (2003)

D. Vilone, C. Castellano, Solution of voter model dynamics on annealed small-world networks. Phys. Rev. E **69**(1), 016109 (2004)

G. Weisbuch, G. Deffuant, F. Amblard, J.-P. Nadal, Meet, discuss, and segregate! Complexity **7**, 55–63 (2002). https://doi.org/10.1002/cplx.10031

H. Xia, H. Wang, Z. Xuan, Opinion dynamics: a multidisciplinary review and perspective on future research. Int. J. Knowl. Syst. Sci. (IJKSS) **2**(4), 72–91 (2011)

Q. Zha, G. Kou, H. Zhang, H. Liang, X. Chen, C.C. Li, Y. Dong, Opinion dynamics in finance and business: a literature review and research opportunities. Financ. Innov. **6**(1), 1–22 (2020)

Luca Marconi is currently PhD Candidate in Computer Science at the University of Milano-Bicocca, in the research areas of Artificial Intelligence and Decision Systems, as well as Business Strategist and AI Researcher for an AI company in Milan. He also gained experience as Research Advisor and Consultant for a well-known media intelligence and financial communication private consultant in Milan. He holds a master of science and a bachelor of science degrees in Management Engineering, from the Polytechnic University of Milan, and a master of science degree in Physics of Complex Systems, from the Institute of Cross-Disciplinary Physics and Complex Systems of the University of the Balearic Islands (IFISC UIB-CSIC). He also holds a postgraduate research diploma from the Advanced School in Artificial Intelligence, Brain, Mind and Society, organized by the Institute of Cognitive Sciences and Technologies of the Italian National Research Council (ISTC-CNR), where he developed a research project in the Computational Social Science area, in collaboration with the Laboratory of Agent Based Social Simulation of the ISTC-CNR. His research interests are in the fields of artificial intelligence, cognitive science, social systems dynamics, complex systems and management science.